THE TREE FARM

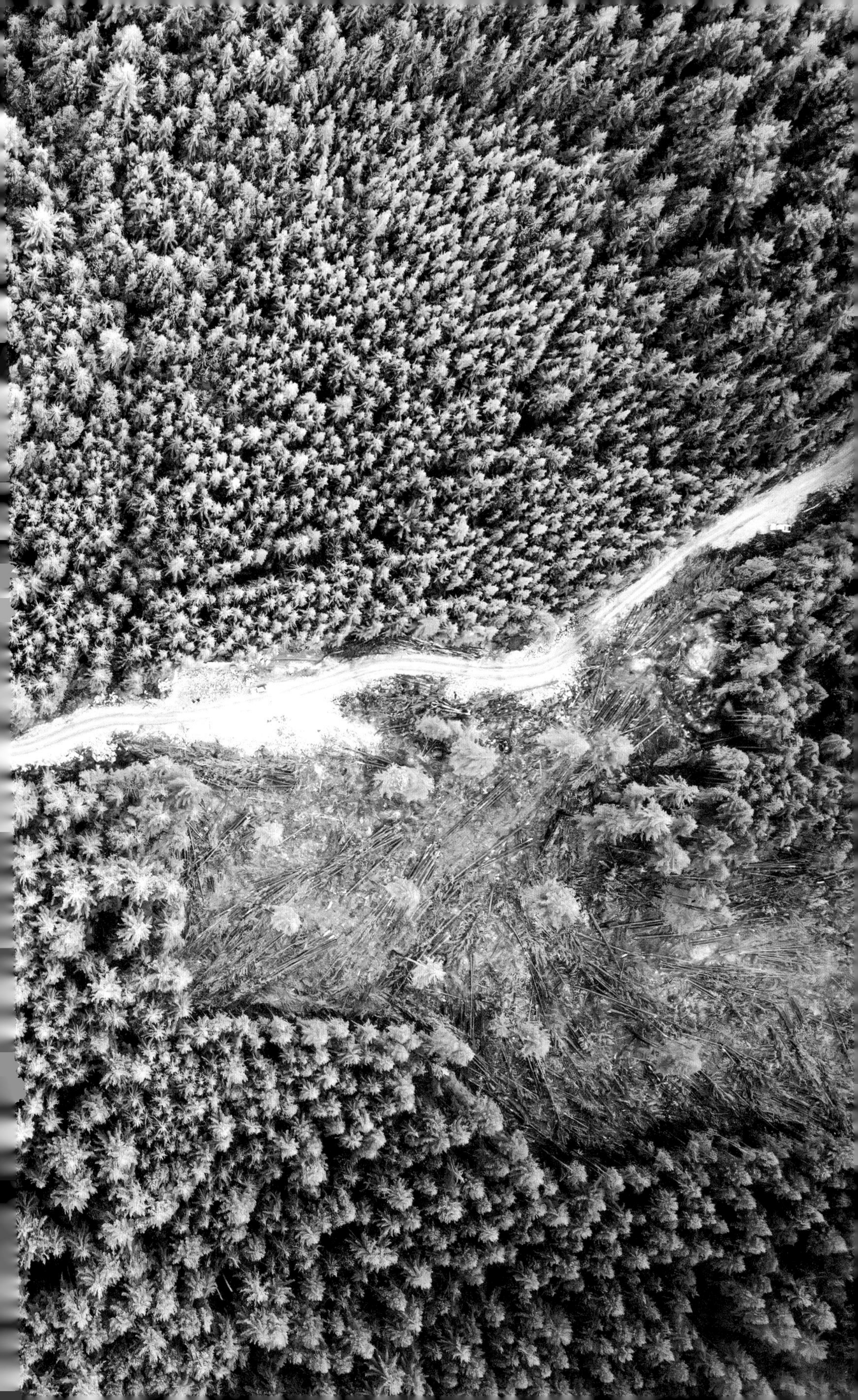

THE TREE

MICHELLE RHODES

FARM

THE EVOLUTION OF CANADA'S FIRST COMMUNITY FOREST

PAGE TWO

Cataloguing in publication information is available from Library and Archives Canada.
ISBN 978-1-989025-68-0 (paperback)
ISBN 978-1-77458-037-0 (ebook)

Page Two
pagetwo.com

Edited by Amanda Lewis and Tilman Lewis
Cover and interior design by Taysia Louie
Cover, frontispiece, and endpiece photos by Jason Brawn
Maps by Eric Leinberger
Illustrations by Emily Gauthier
Printed and bound in Canada by Friesens
Distributed in Canada by Raincoast Books
Distributed in the US and internationally by Macmillan

21 22 23 24 25 5 4 3 2 1

In June 2021, as the book was going to press, District of Mission officially became the City of Mission. References to the District of Mission were retained for historical accuracy.

To Bob O'Neal, without whose support and advocacy this book would never have happened, and to the Indigenous leaders and community stakeholders shaping the future of the Mission Municipal Forest

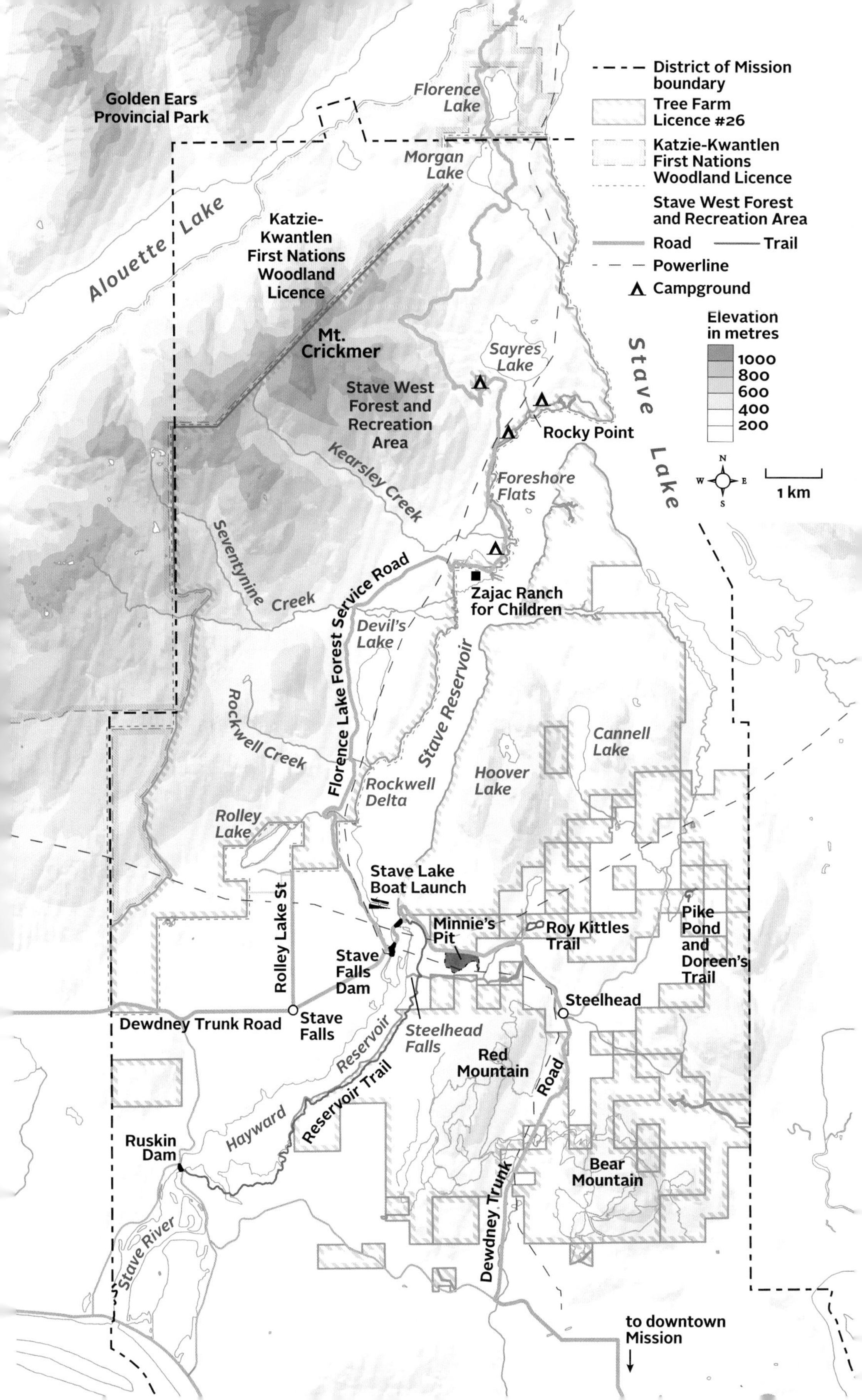
District of Mission boundary
Tree Farm Licence #26
Katzie-Kwantlen First Nations Woodland Licence
Stave West Forest and Recreation Area
Road
Trail
Powerline
Campground
Elevation in metres
1000
800
600
400
200
N
W
E
S
1 km
Golden Ears Provincial Park
Florence Lake
Morgan Lake
Alouette Lake
Katzie-Kwantlen First Nations Woodland Licence
Mt. Crickmer
Stave West Forest and Recreation Area
Sayres Lake
Rocky Point
Stave Lake
Kearsley Creek
Foreshore Flats
Seventynine Creek
Florence Lake Forest Service Road
Zajac Ranch for Children
Devil's Lake
Stave Reservoir
Rockwell Creek
Cannell Lake
Hoover Lake
Rockwell Delta
Rolley Lake
Stave Lake Boat Launch
Minnie's Pit
Roy Kittles Trail
Pike Pond and Doreen's Trail
Rolley Lake St
Stave Falls Dam
Steelhead
Dewdney Trunk Road
Stave Falls
Reservoir
Steelhead Falls
Red Mountain
Road
Reservoir Trail
Hayward
Ruskin Dam
Bear Mountain
Dewdney Trunk
Stave River
to downtown Mission

CONTENTS

Introduction
A WATERSHED MOMENT

IN LATE SEPTEMBER and early October, the rains come for what can feel like an eternity but is usually just six months. The rain's arrival marks the turning point between the two main seasons of southwestern British Columbia's mild climate. The late-summer, sun-soaked warmth gives way to cool, humid conditions that change little from day to day or from day to night. A claustrophobic greyness of low-hanging clouds and sky descends upon mountaintops and the dense coastal forests below.

The novelist Tom Robbins once wrote that October on the Northwest Coast feels like "a wet rag on a salad."

PRECEDING Jumping into Stave Lake on a hot August day. *Jason Brawn*

FACING Mist decends upon the Mission Tree Farm on a cool October day. *Romana Osbourne · all rights reserved*

The rains recharge life on the hills. The forest erupts out of its summertime doldrums. Moss suspended from low-lying branches drinks in the autumnal mist. Quilts of fir branches and hemlock boughs intercept the rain as it trickles to the plush carpet of sodden needle mulch and ferns atop the forest floor.

The millions of cedar, hemlock, and Douglas fir trees blanketing the Mission Tree Farm rely on these persistent seasonal rains. Nestled in the mountains seventy kilometres east of the Salish Sea, the forest thrives in this meeting of the mild, wet coastal climate with the more variable seasonal fluctuations of the montane zones. Most of the moisture falls as rain between November and March; at higher elevations and in colder years, it arrives as snow.

Unlike the lodgepole pines and spruce that cover millions of hectares in the central parts of the province, western red cedar and hemlock are not fire-hardy. Usually, they do not need to be—damp soils and thirst-quenched vegetation burn slowly. When the rains do not fall as expected, or when the mountaintop snows melt early, however, a growing fire danger puts local officials and forestry crews on alert. Never have the timing and duration of these rainy cycles been more important than they are now, when tens of thousands of people annually visit the Mission Tree Farm to hike, bike, pitch tents, park RVs, and light campfires.

FACING Playing near one of the many rain-fed creeks on the Mission Tree Farm.
Kelly Cameron

An Advantageous Location

THE MISSION TREE Farm occupies a middle space between mountain and river, creating a buffer between wilderness and the city. To the north, the jagged, snow-capped Mt. Robie Reid (elev. 2,095 metres) and Mt. Judge Howay (elev. 2,262 metres) command the Mission skyline; both lie just outside municipal boundaries in Golden Ears Provincial Park. The tree farm is divided nearly in half by a hydroelectric complex. It includes the Stave Falls and Ruskin Dams, the Hayward Reservoir, and Stave Lake and Reservoir—whose levels fluctuate widely over the year as water is drawn down to meet demand for power in Vancouver. All of these bodies of water drain to the Fraser. Several small and mid-sized lakes dot the landscape, including Devils, Sayres, Morgan, Hoover, and Cannell, although each of them pales in size compared to Stave.

Mission's tree farm, among BC's smallest community forests, is situated in rugged terrain. Almost all of the forest is second and third growth, much of it having first been widely logged in the early 1900s. But Mission benefits from being in one of the best possible locations in which to operate. Some of the province's most productive forests grow here, and Mission harvests and replants high-value species like cedar and Douglas fir. Mission's proximity to the Fraser River, mills, and markets keeps transportation costs low. In past years, the city has worked with contractors to find high-value niche markets for its wood.

FACING A patchwork of recently harvested areas and immature forest on the Mission Tree Farm, 2012. *City of Mission*

Today, the TFL's neighbours include private landowners, gravel pit operators, BC Hydro, the province, and the Katzie-Kwantlen (K&K) First Nations Woodland Licence area.

As of 2020, the tree farm included 10,935 hectares of land, 88 percent of which is owned by the province. The city owns the rest. Most of the municipal holdings are found in the southern, lower-elevation sections of the forest, near the rural neighbourhoods of Steelhead and Stave Falls. The Mission forest today is maturing but not pristine. Little of it is old growth. This is very much a working forest.

The province limited the original licence to operate a tree farm to include only those Crown forest lands within District of Mission boundaries that did not already have licences on them. This arrangement has remained unchanged since 1958. But Tree Farm Licence (TFL) 26 has expanded multiple times over the years, through municipal land purchases and acquisitions from the province. The most recent changes to the tree farm's boundaries came in 2019, after the District and the province negotiated a land exchange. The deal transferred control of 360 hectares of land along the western perimeter of the tree farm to a new First Nations woodland licence (FNWL). In exchange, Mission received control over

Changing colours in early autumn along the Hoover Lake forest road. *Justine Robinson · all rights reserved*

several Crown-owned timber berths that were previously surrounded by but not included in Mission's licence.

Today, the TFL's neighbours include private landowners, Zajac Ranch, gravel pit operators, BC Hydro, the province, and the Katzie-Kwantlen (K&K) First Nations Woodland Licence area. The K&K is a joint venture between the Kwantlen and Katzie First Nations operating on Blue Mountain in Maple Ridge. The Kwantlen First Nation and Mission have committed to joint ventures when feasible on the adjacent tenures (or licence areas), including training and public outreach.

Stave West planning team members and stakeholders, gathering after a paddle on the western shore of the Stave Reservoir, August 2018. *Jason Brawn*

A Community Ahead of Its Time

THE DISTRICT OF Mission has operated a community forest since 1948, when local voters said yes to creating a forest reserve out of municipal lands.

In the 1940s and 1950s, Mission saw its local industry put at risk by large-scale harvesting carried out by the powerful companies that had locked up access to most of the timber in southwestern BC. Rather than tying its economic fate to these large operators, the District sought to secure a steady, perennial supply of wood. Not satisfied that the establishment of a forest reserve would be enough to support local economic development, Mission applied for its provincial tree farm licence only a few years later. In 1958, the District obtained a twenty-five-year renewable tree farm licence and, in doing so, gained control over most Crown forest lands within municipal boundaries.

It was a bold strategy. No other town in Canada had established a working forest operation on Crown lands. It would be another thirty years before even another BC community would do so.

TFL 26 in hand, the District assumed management over this mix of Crown and municipally owned lands along the southern reaches of Stave Lake, most of which had been heavily logged prior to the 1930s. To make the operation viable, Mission needed to transform the badly overgrown, weedy, and fire-prone second growth into commercially valuable timber stands. District crews and contractors planted millions of Douglas fir, cedar, larch, and pine seedlings. Staff conducted trials in broadcast burning and experimented with new ways of cultivating yellow cedar. Crews laid out roads, installed bridges and culverts, and found ways to sustainably extract logs from steeper, more remote, and higher terrain.

In recent decades, Mission's focus has shifted in part to enhancing opportunities for recreation in the forest. An expanding trail network has been put in place through partnerships with local community organizations. Inmates from local corrections institutions previously helped cut and clear routes into the forest. From 2013 to 2021, the province, District, and Kwantlen First Nation built more than two hundred campsites on the western half of the tree farm. Before that point, there had been none.

Piling recently harvested trees in preparation to load them onto a truck, 2019. *Jason Brawn*

It was a bold strategy. No other town in Canada had established a working forest operation on Crown lands.

The Mission Tree Farm evolved from an experiment in local resource control to what it is today, a community forest that is both ecologically and financially sustainable, balancing logging and recreation and the pressures of population growth nearby. The District of Mission grew as well, from a small, resource-dependent town of a few thousand people in the 1950s to a bedroom community of more than forty thousand residents and counting. Each passing year draws Vancouver's metropolitan influence closer to the community and its forest.

The District responded to population and economic pressures in part by engaging in longer-term planning, but it has not done so alone. The Matsqui, Kwantlen, and Leq'á:mel First Nations have historic and unceded claims to lands and waters of the Stave and share in its cultural heritage. Since 1997, they and other Nations that have asserted claims to the watershed have also had the legal right to be consulted on harvesting and land-use decisions on the tree farm.

Consultation has grown into collaboration, a process that has required time, patience, and trust building between the Nations and the municipal government. This journey remains an incomplete process, but one that Mission and the Kwantlen, Matsqui, and Leq'á:mel First Nations have committed to. One major step has been in shared governance. The three Nations and the municipality work together on recreation planning and land-use decision making for the Stave West Forest and Recreation Area (SWFRA), the western half of the tree farm.

A Tree Farm by Any Other Name

"TREE FARM" CAN be misleading. The name evokes images of a U-cut Christmas tree producer, with long rows of evenly spaced and aged noble and balsam firs. The terminology stems from the type of licence the provincial government awarded to the District. At the end of World War II, when BC created the modern tree farm licensing system, the goal was to grow as many trees as possible to support a rapidly expanding timber industry. Licensees would quite literally farm trees.

In 1990, the District's forestry office changed the name of the tree farm to the Mission Municipal Forest, although there was no change to the language used in the licence itself. For that reason, city staff and members of the public can be heard using either name, or both interchangeably. At the time, the phrase "community forest" was not yet commonly used. Today, however, forestry staff frequently introduce the operation as Mission's community forest.

The legal and political language used to describe the forest masks the traditional and historic names for the Stave and other places within the boundaries of the tenure. The Coast Salish and their ancestors have lived continuously in this corner of BC for close to thirteen thousand years. The Halq'eméylem language was born just to the east of here, and from it, thousands of names have been given to important meeting sites, natural features, villages, and resource sites. These places were part of S'olh

The logo and Halq'eméylem name given to Stave West, shown here on the hat Payne is wearing. Stave West is the western half of TFL 26. *Terry Hood*

The reservoir created by the damming of the Stave River flooded areas used by Coast Salish Peoples over many generations. *Kelly Cameron*

Temexw, the traditional territory of the people of the river, or Stó:lō. Settlers renamed this river after the explorer and trader Simon Fraser, who arrived in the lowlands in 1808.

In the Stave Valley and nearby Alouette Valley and surrounding regions, families and communities carved canoes, caught salmon, travelled and traded and created maps absent fixed boundaries, and they celebrated, held Potlatches, feasted, and struggled together. The Peoples of the Stave developed a deep, living knowledge of the rivers, valleys, and uplands of this region, rooted in centuries of settlement, seasonal movements, ritual, and thrift.

Colonization replaced Indigenous toponomy with settler place names, and the original names given to Stave Lake and River were lost to disease and displacement. Collective and cultural memories and practices remain anchored to this place, however, along with a desire to name in Halq'eméylem features such as trails on the tree farm. The Kwantlen First Nation has bestowed a Halq'eméylem name upon the lands west of Stave, one that honours the common values and rich culture of the area:

mekw'wa't a'xwest ikw'elo'

Everyone shares here

A Watershed Moment

EVERYONE SHARES HERE. What happens, though, when everyone shows up all at once?

In late March 2020, BC's health officials, concerned about the possible spread of a highly infectious new coronavirus in public spaces during the COVID-19 pandemic, ordered the temporary closure of provincial parks and BC Hydro–run recreation sites. Local governments followed suit, closing the smaller regional parks and public facilities. Eat-in restaurants, theatres, and concert venues shuttered their doors. The federal government shut the US border to non-commercial travel, and people were told to stay close to home.

The timing of the closures coincided with the arrival of an unusually warm and dry April—one of the driest on record, in fact, for the south coast. Cooped up and craving diversion, residents of Vancouver's suburbs headed east towards Mission, past the gated and barricaded turnoffs for parks, boat launches, and picnic sites. They checked social media accounts and crowdsourced trail apps to find new places to hike and bike. Visitors congregated at popular areas like Bear and Red Mountains but extended their search outwards as well. Nearby, drivers jockeyed for space at the end of the normally tranquil dead-end Sabo Street, all for the pleasure of slogging up a moderately steep, dusty, 4.5-kilometre-long logging road to a scenic lookout.

Mission has not only endured but has **become a role model of sustainable community forestry in Canada.**

Logs harvested in the morning, to be loaded and hauled down the mountain in the afternoon. *Jason Brawn*

The COVID-19 pandemic also temporarily affected the tree farm's bottom line. The resulting economic slowdown pummelled an industry already struggling in recent years with low log prices and trade uncertainties. This was the latest in a string of widespread and persistent crises in BC's forest economy since the late 1970s, crises that have at times begged the question of whether small operations like Mission's could survive.

The Tree Farm reveals the reasons why Mission has not only endured but has become a role model of sustainable community forestry in Canada. Two factors stand out. First, the tree farm has succeeded because it has been led by effective, visionary leaders who were willing to take risks and prioritize community needs. The District's early forestry professionals were self-taught in the art of managing a forest for local rather than corporate interests, at a time when the provincial forest economy and tenure system (licensing) favoured large forest companies. Today, shaping a vision for Mission's forest is a shared process, drawing on the talents, tenacity, and commitment of city staff, Indigenous leadership, and community groups.

The second factor has been the ability to see past short-term adversities towards long-term solutions. When Mission took on TFL 26 in 1958, the poor condition of the land required the District to redirect revenues from timber sales towards restoring and replanting the forest and building better roads. The decision to take on a project that would not generate funds for the District's general coffers for nearly a decade is remarkable for any era. In the 1980s, financial crises led to a streamlining of operations and the creation of rainy-day funds. And by the 2000s, as undesirable, even criminal, activity turned parts of the tree farm into a "wild west," Mission and neighbouring First Nations began to work on revitalizing the area now known as Stave West.

This book is born out of a watershed moment in the history of Mission's tree farm. Population growth and surging visitation will reshape the tree farm and its relationship with the surrounding community in the decades to come. So, too, will a warming climate and continued turbulence in the logging industry. Future residents and visitors are more likely to appreciate the community's forest for recreation and education than for managed, even if small-scale, logging.

The qualities required to sustain Mission's forest in the future are those exemplified by the people and stories told in this book: foresight, leadership, investment, resilience, adaptation, and collaboration. This last one is key. While the tree farm requires support from locals and elected leaders, it does not "belong" to Mission alone.

Forest planning today more directly engages a network of organizations who also use the space. The "community" in community forestry includes forest workers and recreation users who come to Mission from across southwestern BC. At the table as well are those for whom the Stave watershed is a traditional, spiritual, and shared territory. Collectively, these diverse voices are helping to shape the future of a community's forest in Mission.

Mission Forestry's Department, July 2019. Back row (L to R): forestry operations' Pat Watson and Brett Saniger, director of forestry Bob O'Neal (ret.), foreman Brad Laughlin, current director Chris Gruenwald. Front row: forest technologist Kelly Cameron, manager of forest business Dave Heyes (ret.), forestry operations' George Kocsis, administrative assistants Michelle Weisgerber and Erika Duplisse. *Jason Brawn*

SEEDS OF AN IDEA 1

THE DISTRICT OF Mission was an early leader and innovator in Canadian community forestry. For more than a century and especially after World War II, large timber companies sustained most of rural British Columbia and the provincial economy. Mission, meanwhile, foresaw a future where the fortunes of this growth passed them by. The town took a calculated risk on community management, a move that helped support a local industry even when, by the late 1980s, timber-dependent communities elsewhere were in free fall.

In the first 30 years after Mission received its tree farm licence, only a few other BC communities—easily counted on one hand—followed suit. Since the late 1990s, however,

PRECEDING New growth. *Romana Osbourne · all rights reserved*

FACING Remnants of early logging can still be found on parts of the tree farm. *Kelly Cameron*

a growing number of BC towns and First Nations have pursued community tenures of their own under new licensing systems and agreements. Many of these newer licences were established in the aftermath of major changes to the industry: consolidation, overharvesting, global competition, and a public desire to see more protections for old-growth forests.

Just as these newer operations can draw from Mission's experience, Mission, too, is learning from them—about differences in organizational structure and the ways in which First Nations partnerships are developed. These professional networks could not blossom soon enough. With each passing year, local forest managers are asked to plan for more, and more complex, demands on the forests.

Director of Forestry Bob O'Neal leading BCIT students in a training exercise at Rockwell Delta, 2019. *Terry Hood*

The Argument for Community Forests

AROUND THE WORLD, community forests take on many forms, legally and operationally. Community forests are found in wealthy countries and poorer ones. Canada has a growing number of community tenures. Despite a population ten times the size of its northern neighbour, the United States has relatively few. Even within Canada, the number and types of community forests vary among provinces and territories.

At the heart of any community-managed venture is a conviction that local populations are usually better equipped to make decisions on how to use and sustain forest resources for long-term benefit. Local residents, after all, bear the impacts of activities occurring in nearby forests. Most community forests in North America operate on land owned by local, provincial or state, or regional governments rather than on federally owned land.

In community-managed forests, residents exercise a higher degree of control in how the forest's resources are used. In practice, this can mean anything from obtaining ownership of the land to full participation in day-to-day activities that involve the forest. Few residents are trained in the technical aspects of the industry, but they possess other forms of valuable knowledge that come from living in or next door to a forest.

People who have resided in an area for some time usually can claim cultural, spiritual, economic, and personal ties to the forest. Where people still practise strong cultural traditions, such as berry picking or seasonal hunts, many members of the community may share in stewarding forest resources.

Community forestry in Canada took inspiration, at least initially, from systems in northern Europe. Mission's political leaders looked to the Swedish experience in the 1940s. As might be expected from the country that gave the world IKEA, Sweden manages its community forests for maximum efficiency and minimal waste. Day-to-day management is left to forestry professionals, who emphasize producing trees for market rather than ecological diversity.

Most community forests are located close to the communities that manage them. Having local control over the forests means that decision makers should, in most cases, account for the diverse values and goals that are important to the community. These include demand for recreational amenities like trails and boat docks but also education and interpretation.

Community forests lead in protecting and providing for cultural needs and values and in efforts towards reconciliation with Indigenous Peoples. A local forest operation may work with the community and businesses to develop their own mills or tourism products. Community forests play a critical role in reducing the risk of urban-wildland interface fires.

In Canada and other Western countries, communities often turn to local management of their forests as one way to hedge against difficult times in the timber industry. Large companies not rooted in a community will respond differently to a downturn or a recession than a local government or non-profit board will. Multinational firms may shift their resources to more profitable mills elsewhere, shrinking jobs within a community. They are also less likely to prioritize spending on amenities, such as trails for area populations, or maintaining local employment opportunities.

Community forests provide economic benefits, such as funnelling timber to local mills or offsetting some of a town's tax burden. And then there are those harder-to-measure qualities, such as scenery and wildlife protection. While one cannot, as the saying goes, eat scenery, residents may enjoy higher property values by virtue of living next door to it.

At the heart of any community-managed venture is a conviction that local populations are usually better equipped **to make decisions on how to use and sustain forest resources for long-term benefit.**

Community Forestry in British Columbia

MISSION WAS THE first operational community forest in BC and the second community to set forest lands aside for a local forest reserve. The Municipality of North Cowichan on Vancouver Island and the District of Mission established municipal forest reserves two years apart, in 1946 and 1948, respectively. In both cases, the land earmarked for the reserves was owned outright by the communities. Mission began working the reserve almost immediately. Within a year of its creation, the District had begun a timber inventory and started replanting denuded lands with more marketable species. North Cowichan did not begin forest operations until the early 1960s.

Mission's forest operation is also the first in Canada to be run by a single community. Ontario introduced projects in local forest management as early as 1912, but these projects were managed at the county level. Whether this is a significant difference or not could be debated; it depends how large an area one considers to be a community.

Regardless, community forests were rare in BC and Canada up until the late 1990s. Mission and North Cowichan operated the only two community-managed tenures in the province until 1993. That year, Revelstoke obtained a tree farm licence, the same type of licence to operate that Mission has.

FACING The family of the late Mission logger and silviculturalist Roy Kittles gathers around the tree farm's oldest Douglas fir, near Steelhead. *Kelly Cameron*

Other communities had expressed an interest in developing their own local forest ventures. At the eastern end of the Fraser Valley, the Village of Hope suggested as much to a 1955 provincial commission examining the sustainability of BC's forestry sector. Unfortunately, the village lacked the support of the local board of trade. It would have to wait another fifty years until Hope, the Yale First Nation, and the Fraser Valley Regional District were issued a shared licence for the Cascade Lower Canyon Community Forest.

In the early 1970s, representatives of communities in BC's Interior (including Williams Lake, Burns Lake, and 100 Mile House) and Vancouver Island toured the Mission Tree Farm. These communities lobbied the provincial government for more tree farm licences to be distributed to their municipalities. The licences did not materialize, likely due to a lack of political will in Victoria and the challenges of finding timberlands not already locked up in other leases issued to private operators. Residents in the Slocan Valley and Smithers-Kispiox Valley also pushed unsuccessfully for the province to create new types of licences for community-managed forests. Kaslo and Creston did eventually procure fifteen-year volume-based licences that gave them each access to a set quantity of harvestable timber on nearby Crown land.

The nature of land ownership partially explains the inability to develop more municipal operations sooner. In BC and the rest of western Canada, more than 90 percent of forest lands are held by the Crown. This makes forest management a provincial matter. The right to occupy and harvest wood from BC's Crown lands is granted through a tenure system. This involves the allocation of licences for woodlots, timber sales, tree farms, and, in recent years, dedicated community forest agreements (CFAS).

Forest planning in BC is organized around how much timber can be sustainably harvested. The provincial tenure system identifies where to cut trees, how many to cut and at what rate, who will oversee the cutting, and how much they will pay for every cubic metre they do cut (known as stumpage). There are rules around how to plan the cut, how to build roads to what will be cut, how to replace what has been cut, and how to salvage trees that would otherwise rot before they could be cut.

These rules are absolutely necessary to ensure that BC's forest sector can operate in a transparent, efficient, and accountable way within a global marketplace. That said, BC's forests lack the orderly precision of managed Scandinavian forests, and with good reason. Planting long rows of identical trees of the same age is not the same thing as recreating a forest.

Across BC, the public's opportunity to engage in forest management is limited. Management and stewardship plans do involve a formal public consultation period. Understanding forest plans and reports, however, requires at least some working knowledge of forest practices. Even people living near forests could not define what a basal retention area is or determine whether the proposed stocking standards are too high.

Most of BC's productive forest land is allocated to medium- and large-sized firms through renewable tree farm licences. Each renewable licence was typically issued for up to twenty-five years. When Mission took on TFL 26 in 1958, it did so knowing that it would have to follow the same procedures and regulations as the big timber companies. The one exception was stumpage. Until 1979, the District paid annual royalties rather than the higher stumpage rates, at a significant cost savings for Mission. For most other needs, the District hired seasoned forest engineers and registered professional foresters to manage the day-to-day operation.

Even in this regulated, professionalized model of forest management, community forests deviate from business as usual in how they are governed. In BC's community forests, decision making usually is undertaken by one of four types of governance structures: a special-purpose corporation made up of local government representation; a multi-government body, with representatives from First Nations and local governments; an appointed board; or a municipal forestry department answerable to local government. This last type, the structure that Mission employs, is the least common.

Growing Out of Crisis

MISSION'S MUNICIPAL FOREST is unique even when compared to other community forests in British Columbia. The tree farm operates under a different type of licence than most, it encompasses forest lands only within municipal boundaries, and it has historically been managed by a branch of the municipal government.

These differences can be chalked up to how much time it took for the second community forest, then the third and the fourth and so on, to be established. Almost all of BC's community forests date back to 1998 or later, with many of them established in the last decade. These newer tenures share more in common with one another than with Mission.

Despite these differences, Mission shares a common cause with many other BC timber towns. The District sought out a tree farm licence in the 1950s in large part because its local wood products industry was in trouble. The same was true for many communities that would later apply to manage their own forests. Mission just felt the industry's changes earlier than most.

In the 1950s and 1960s, BC's timber industry was booming. Technological change allowed forest companies to harvest at faster rates and higher volumes. Companies built massive mill complexes that produced plywood, dimension lumber, and pulp. Well-paid jobs in manufacturing and the forests were common. For timber-dependent towns in much of BC, it would have been hard to imagine that the large pulp and lumber mills built in the 1950s and 1960s would ever become obsolete.

If there is one reason why BC's communities did not share Mission's aspirations for community forestry during the postwar years, this was it. The security of work in BC's forests and mills allowed local communities to rely on the tax revenues from timber companies and their workers. The companies applied for tree farm licences and kept workers on the payroll for decades. Social scientists refer to this time as the Fordist era, when the mass production of logs and dimension timber rivalled the scale and efficiencies of automobile factories.

The large mills in southwestern BC that sprang up during and after World War II consumed timber from the Fraser Valley. Mission's more modestly sized mills struggled to compete. As early as the 1950s, Mission knew that its locally owned mills would have problems getting enough timber. TFL 26 provided a partial remedy for this challenge.

By the late 1970s, the industry faced a series of blows that left communities reeling. Oil shocks, foreign competition, trade disputes, and technological change reduced demand for workers. Harvests exceeded sustainable levels, and mills struggled for supply. Environmental activists blockaded roads and demanded tenure reform. In the 1990s, the province loosened requirements that timber be milled close to the source. This change provided forest companies with some economic relief but also made them footloose. Less profitable mills closed, and companies were broken up, sold, or restructured. By the end of the 1990s, MacMillan Bloedel, once the largest timber company in BC and one of the largest in the world, was no more.

Small cities and regional governments lobbied for greater control over local resources, hoping to stem the tide of job losses and improve long-term sustainability. Some communities were motivated to expand forest-based tourism and outdoor recreation opportunities and to protect habitat and water quality.

As part of a package of reforms designed to revitalize struggling timber communities, BC's government, led by the New Democratic Party

(NDP), piloted a community forest program in 1997. The following year, it amended the Forest Act to allow for community forest agreements, and seven communities were awarded provisional licences. Like a TFL, a CFA is a renewable, area-based licence, providing communities with long-term access to forest lands nearby. Within the decade, the province expanded the program and the number of licensees.

Since the introduction of CFAS, more than one hundred communities have taken advantage of them, creating more than five dozen distinct community forest operations as of 2021. Community forest licence holders are not a one-size-fits-all group. Some CFAS are managed by boards with strong involvement from local residents, while others are managed with less public input and participation. Still other CFAS are overseen by municipal or First Nations governments, and several are managed through intergovernmental partnerships. For now, Mission and Revelstoke remain outliers, continuing operations as tree farm licensees.

Local populations still oversee only a small fraction of the BC's and Canada's forest lands, even as the number of community forest tenures grows each year. Timber companies continue to do most of the harvesting on public lands. Critically, however, local organizations and municipalities are managing an increasing share of the forest within BC's urban-wildland interfaces, close to population and industry.

Indigenous Nations now manage more of BC's forest land through dedicated First Nations woodland licences. Creating new business and training opportunities is particularly important to most First Nations CFA and FNWL holders. One of the newest FNWLS, managed by the Katzie-Kwantlen, abuts the Mission Tree Farm. The city and the Kwantlen partner on recreation-related projects, training, and knowledge sharing. The Kwantlen First Nation also signed a memorandum of understanding with the District of Mission to move towards greater joint management on the tree farm.

Community forests in BC have had a mixed track record of success, financial and political, but most are still young. Mission's operation is not only the longest-running, but it has also been the most stable over

For timber-dependent towns in much of BC, it would have been hard to imagine that the large pulp and lumber mills built in the 1950s would ever become obsolete.

time. The District has also enjoyed a historically advantageous situation. Unlike forests in the BC Interior, Mission has not had to deal with the ravages caused by the pine beetle, stand-clearing fires, or the challenges of getting wood to markets.

Mission's tree farm also benefits from its location in some of BC's most productive forest lands, with high-value timber species such as western red cedar and Douglas fir. Mission's proximity to the US border and the Port of Vancouver helps, as does being close to a large and expanding population that needs wood products.

Coastal forest management comes with added complications, though. Mission's tenure is small compared to most CFAs, with little room to expand outwards. Its proximity to neighbourhoods put Mission's operations under the watchful eye of a growing population that does not want to look out their windows at clearcuts.

In short, when it comes to community forests, geography, as well as history, matters.

2

THE STAVE VALLEY

AT A BIT more than a hundred square kilometres, the Mission Tree Farm is a small fraction of the size—less than 6 percent—of BC's newest community forest licence, which was awarded to Fort Nelson in 2019. The tree farm and Fort Nelson Community Forest could not be more different. Mission benefits from highly productive coastal forests with high-value species, while Fort Nelson will be harvesting boreal forests with smaller-diameter trees used for pulp and timber. Located in the lower Stave Valley, Mission's topography is challenging and requires harvesting in small cutblocks. Fort Nelson's relatively flat terrain will allow it to maximize its economies of scale in logging. Whereas Fort Nelson is a town still dependent on the

PRECEDING Rocky Point. *Jason Brawn*

FACING The size of Mission's cutblocks, or harvest areas, is limited by terrain, waterways, and the small size of the tree farm licence area. *Jason Brawn*

A three-person crew handles most of the harvesting on this cutblock west of Stave Lake. *Jason Brawn*

timber industry, Mission manages the tree farm on the outskirts of a rapidly growing metropolitan area.

Each of British Columbia's community forests operates in a unique geography, defined by its terrain, geology, hydrology, soil, climate, ecology, and cultural and economic history. Harvesting and forest management practices that work well in one corner of the province may not work in another.

The Stave Valley's glacial history, undulating terrain, and mild, temperate climate allowed for the evolution of a richly diverse forest ecosystem. Glaciers carved out the deep, fjord-like shape of Stave Lake, perfect for power production and power boats. More than a century of resource production and development, however, has fragmented and simplified the ecology of the lower Stave, and climate change further exacerbates the risk of losing iconic species like yellow cedar. It may be a small tenure by comparison, but the Mission Tree Farm will have a big role to play in working with First Nations, community, and government agencies to conserve and enhance the diversity of this ever-changing forest landscape.

Rock and Ice

THE STAVE WATERSHED is among the most geologically and archaeologically significant areas along the southern flanks of the Coast Mountains. Thousands of years ago, ice sheets many metres deep repeatedly advanced and retreated along the Stave, scouring the valley's walls and leaving thick deposits of till at the glacial tongue. As the last Ice Age came to an end, the Stave watershed was the first area of the Lower Mainland to be populated. Resting in glacial till and buried by silts and forest soils, the stone tools and carbon deposits from campfires left behind by the first peoples of this place are evidence of the intermingling of human time and geologic processes.

Geologic time scales are understandably difficult to wrap one's head around, given that what is measured—the Earth's age—is more than 4.5 billion years. Each stage in this geochronology is marked and bounded by significant events that are recorded in fossils and sands: the creation of a supercontinent, the first available oxygen, the disappearance of the dinosaurs. Glaciation, plate tectonics, volcanism, the erosive powers of wind and water, and biotic evolution further reshape the Earth's surface.

Understanding the geologic landscape of the Stave requires going back to the late Pleistocene—better known as the Ice Age. During the Pleistocene, most of the northern hemisphere was covered in ice. Glaciers expanded, contracted, and expanded over and over again as the global climate experienced alternating periods of warming and cooling.

Continental ice sheets up to three kilometres thick in places covered most of North America in the late Pleistocene. The largest of these, the Laurentide ice sheet, extended across most of Canada and the northern United States. This glacial event reached its maximum between twenty-four and fourteen thousand YBP (years before present).

In British Columbia, glaciers developed in alpine regions starting approximately thirty thousand years ago, and over the next six thousand years, spread out and merged to form the Cordilleran ice sheet. This latter period of glacial expansion is referred to as the Fraser Glaciation. In the Fraser Valley and in northwest Washington State, glaciers reached their southwestern and westernmost extent by 16,500 YBP.

At one point during the Fraser Glaciation, ice covered the Stave Valley more than two thousand metres deep. Then, in the five thousand years that followed, glaciers intermittently retreated as climates warmed. The Cordilleran ice sheet began to thin and break apart at its edges. Water once trapped in the world's glaciers flowed outwards to the sea, and sea levels rose. By 13,500 YBP, the Stave Valley was once again largely free of ice.

At the same time, although much more slowly, the valley and surrounding mountains experienced a phenomenon called isostatic rebound. Imagine pressing your thumb hard against your skin, leaving an indentation. Once you lift your thumb, your skin slowly resumes its normal shape. Glaciation had this effect on the Earth's crust. When glaciers were at their greatest depth, their unthinkable weight bore down on the surface. Sea levels were higher relative to land, not because the oceans held more water but because of this downward pressure.

Ice sculpted valleys and reshaped mountains. The geographer Michael Church wrote, "While tectonics has established the architecture of the province, the detail has been wrought by Pleistocene glaciation." The ice carved out new lakes and depressions, which left as their legacy the U-shaped valleys found in the Coast Mountains. As glaciers moved down into mountain valleys, the meltwaters below the ice carved deep channels that rivers such as the Stave and Alouette flow through today.

Glaciers advanced and then retreated, the scouring action loosening chunks of surface materials. Most of this rock was granitic, formed during

At one point during the Fraser Glaciation, **ice covered the Stave Valley more than two thousand metres deep.**

The dynamic glacial past of the Stave is visible it the area's jagged peaks and steep, u-shaped valleys. *Kelly Cameron*

periods of great volcanic activity between 100 and 50 million YBP. The till from the ice sheets would be deposited at the edges of the glacier in moraines, mound-like structures of gravel, sand, and other materials.

Under the massive power and weight of the ice, some parts of the Lower Mainland were once as much as three hundred metres lower than they are today. Arms of the sea extended inland, as far east as present-day Haney just to the west of Mission. Seawater carried marine sediments aloft, sediments that were eventually deposited at the water's edge. Silt, shales, sands, and similar materials covered the granite bedrock in the lower Stave area. The geologic profile of the lower Stave captures a time when this area was the meeting point between ice and coastline, when marine sediments mixed with glacial till.

As the Ice Age came to an end, glaciers persisted at the highest elevations but largely disappeared elsewhere. Most peaks in the southern Coast Mountains were free of snow by thirteen thousand YBP. Forests, fish, and people closely followed the retreat of the ice at lower elevations, and they moved into the Stave Valley no later than twelve thousand YBP. Permanent settlement occurred in the Stave even before it did on the Fraser Valley floor, as the Fraser Valley was still largely inundated with seawater and rivers swollen with glacial melt. The lower Stave Valley provided a safe harbour, among the few spaces high enough and dry enough, a space between receding glaciers and receding seas.

People settled in the uplands near the confluence of the Stave and Fraser Rivers almost continuously during the Holocene epoch, the period that followed the Ice Age. For much of the Holocene, warmer and drier conditions prevented the re-emergence of glaciers, except for some that developed during shorter cold periods. Up and down the Coast Mountain Range, glaciers expanded again from between 6,900 to 5,600 YBP, and their retreat was followed by an increase in the size and number of settlements in the Stave region. When the climate cooled again, between 4,000 and 3,500 YBP, settlements shrank.

Glaciers returned to BC's south coast by the sixteenth century during the Little Ice Age, a period of significant cooling followed by the expansion of alpine glaciation. From the highest points in the Coast Mountains, the glaciers expanded outwards into surrounding river valleys. These were visible at lower elevations well into the nineteenth and early twentieth centuries. Tourist lodges, ski resorts, and park promoters put photographs of these glaciers on brochures and postcards. Mining speculators and government cartographers traversed ice fields and glacial lobes, including in the upper Stave Valley, as they ventured deeper and higher into the backcountry.

Glaciers have been gone from lower elevations of the Stave Valley for almost a century. Placer miners during the late nineteenth century reported finding gold in the vicinity of Clearwater and Glacier Creeks above where they flowed into the northwestern corner of Stave Lake.

The creeks originated from a glacier on the eastern slopes of Mt. Baldy—known today as Mt. Robie Reid. Photographs of the area show that the glacier was visible at the lake's edge as late as the mid-1920s.

The glaciers of the Coast Mountains are shrinking in mass and area and have been for the past century. They are losing volume, thinning most dramatically at the edges. This only quickens the ice's recession.

The closest glacier to Mission is the Stave Glacier in southeastern Garibaldi Park. The edge of the glacier is approximately forty kilometres northwest of its namesake lake, and meltwater flows downhill to the Stave River via Tingle Creek. The glacier, which was among the earliest in the Coast Mountains to begin receding, is retreating at a faster rate than most other glaciers in the region. This is unsurprising, given its southern location and lower elevation. The Stave Glacier is only one of more than 150 glaciers in Garibaldi Park, almost all of which have also been shrinking.

Anthropocentric climate change is hastening the retreat of glaciers across BC and around the world. Glaciers have advanced and retreated at many points during the last seventeen thousand years. Most of these changes occurred before the last two hundred years, when human activities began distorting the balance of greenhouse gases in the atmosphere. What are otherwise natural processes have been sped up. Many southwestern BC glaciers will not make it to the end of the twenty-first century. The legacy of the ice, however, will remain readable in the topography and waters of the Stave.

Genetics

ON THE OUTSIDE, the yellow cedar (*Chamaecyparis nootkantensis*) and the western red cedar (*Thuja plicata*) seem like fraternal twins. They share almost everything with one another. These iconic giants prefer many of the same coastal environments. Western red cedar will happily eke out an existence wherever the climate is not too hot and the soils are rich or wet enough. The species extends into dank, low-elevation forests as far east as southeastern British Columbia and northwestern Montana. The slower-growing yellow cedar thrives in cooler temperatures and in mucky, poorly drained soils found at higher elevations in the Pacific Northwest. They are most plentiful in northwestern BC and Alaska.

The western red grows taller, but the yellow is more highly prized; its wood is harder, the inner bark a bit softer. The two trees splay scale-like leaves rather than needles and are clad in deeply grooved bark. Both trees survive, if left undisturbed, for centuries. The people of the river, the Stó:lō, relied for centuries on both species to provide the raw materials needed for survival and trade.

That is where the similarities between these trees end. At the genetic level, differences abound. The two cedars are less like close siblings than they are distant cousins. Both are members in the ancient family of *Cupressaceae* (or cypress), which also includes redwoods, junipers, sequoias, and other types of cedar. But one would have to go back nearly a hundred million years to find a common ancestor or ancestors.

The Stó:lō Peoples and their ancestors settled and shaped forest ecosystems more to their needs, through harvesting of cedar and other resources.

Yellow cedar emerged as a distinct species between forty to seventy-five million years ago. The evolutionary path taken by *C. nootkantensis* has been difficult to trace and has only been made possible through genetic analysis and molecular science. Plant biologists have so far determined that the yellow cedar is first cousin to cypress trees native to the southwestern US and Mexico. In contrast, the western red cedar's closest relatives are in Korea and Japan. Both species share an affinity for the climates of the Northwest Coast.

One of the places where the ranges of yellow and western red cedar historically overlap is Mission's tree farm, in the lower Stave Valley. Despite how ancient each species is, both western red and yellow cedar are relatively new arrivals, appearing ten to twelve thousand years ago after glacial retreat at end of the Ice Age.

The cedars developed their own unique adaptations to their environments over time, which has ensured their survival for thousands of years in the Stave. The Stó:lō Peoples and their ancestors settled and shaped forest ecosystems more to their needs, through harvesting of cedar and other resources. Massive transformation to the shared territory of the yellow and western red cedar would not occur until the introduction of industrial forestry at the end of the nineteenth century.

FACING TOP Western red cedar (*T. plicata*), or pá:yelhp in Halq'emeylem. *Emily Gauthier*

FACING BOTTOM The distinct shape—and odour—of ch'ókw'e, or skunk cabbage (*L. americanus*) makes its appearance in swampy areas of the forest floor in early spring. *Emily Gauthier*

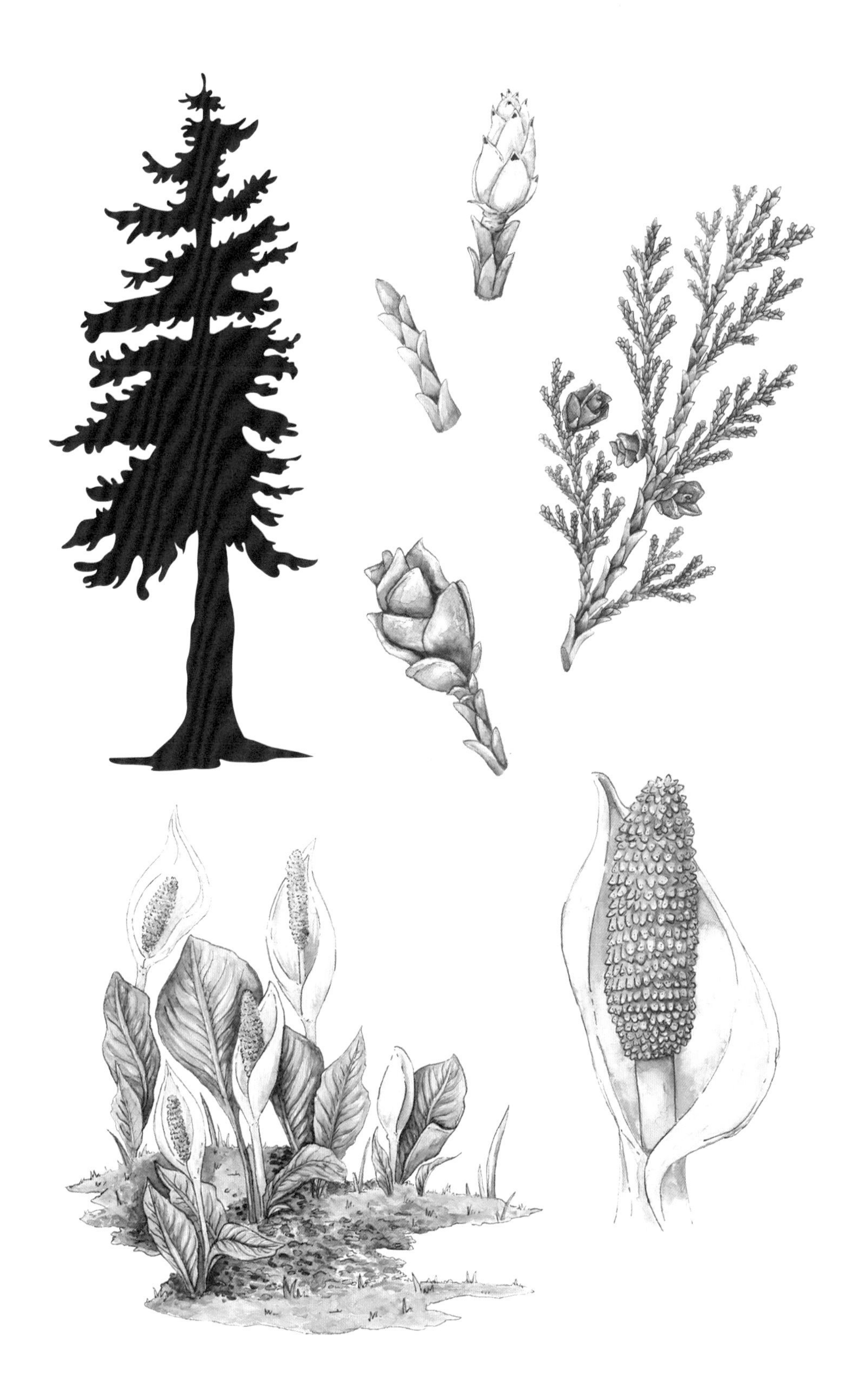

Abundance

ON THE NORTH shore of the Fraser Valley, a steady rain affords unflagging companionship in the fall and winter months. Moisture remains aloft in a dome of leaden clouds that blurs time and space. Every living thing in this coastal rainforest relies on the surety of this rain.

The arrival of the rainy season brings a close to the busy summer recreation season on the Mission Tree Farm. By October, wet weather and an evanescent afternoon light leave less time to get outdoors. For the most part, only locals are on the trails during the week. The preferred routes are sheltered from the rain and snow by dense forest canopy—a short walk down to Steelhead Falls, an after-work bike ride up the west side of Bear Mountain. Up high, walls of cloud erase the mountains and valleys from far-off vistas.

Over the course of any given year, between 1.8 and 2.3 metres of precipitation will fall on the Mission forest. Most of it falls as rain. Snows are more frequent and persist longer at higher elevations. November is the wettest month of the year with, on average, four times as much rain as falls in July. Variable terrain affects precipitation patterns in unexpected ways. West of Stave Lake, precipitation in some form will fall on two out of every three days in the month. One out of three days will receive more than five millimetres of rain or snow. On the west side of the forest, July is the driest month, and on the east side, August is.

FACING Commonly spotted in the forest is the t'ege'lhp, or salal (*G. shallon*), shown here with t'áqe, salal berry. *Emily Gauthier*

• • •

The Roy Kittles Trail in Steelhead cuts through a mixture of mature and younger forest and provides a window into the forest's ecological diversity. Among the more common shrubs is the dark green, waxy-leafed t'ege'lhp, or salal. The Stó:lō used t'ege'lhp and its berries, t'áqe, as a food source, and here, this low-lying bush creates hardy ground cover beneath the conifers. Pacific trailing blackberry creeps along the trail's and road's edge, over mats of curling leaf litter. Sword (slháwél), licorice (st'uslóye), deer, and goldback ferns intermingle on the slopes and in the muddy bottoms between the trail and creek.

Even more impressive are the mosses, lichen, and fungus that blanket the forest in a painter's palette of vibrant green and yellow hues. A springtime walk reveals stepmoss, fan moss, fork moss, yellow moss, Lyell's bristle moss, badge moss, juniper moss, hairy screw moss, golden moss, cat-tail moss, Douglas' neckera, and the delightfully named slender beaked moss. Some of these mosses live on specific trees, while others move onto any available real estate. All of them thrive where only a paucity of light breaks through.

In such a wet climate, detritus decays quickly, helped along by numerous species of fungi. The red-banded and horse hoof fungi are most recognizable, distinguished for their large size, colourful striations, and shelf-like structure. These half-moon fungi perch on the trunks (dead or alive) of birch trees and conifers. Bird's nest mushrooms and orange and brown jelly fungi colonize the bark of fallen trees, alongside gilled and fir cone mushrooms of different types, shapes, and sizes.

Young red alder (é:yth'elhp) and hemlock (mélmélhp) crowd in along the upper reaches of the trail, depriving the forest floor of light. Winter takes its toll, with heavy rains and snows or high winds. Trees rooted in the loose soils and rotting nurse logs will lean or bend over, some until they are almost parallel to the forest floor. By late spring, a few will have toppled over altogether. Mission's forestry crew comes out with saws to clear the trail.

FACING One of the forest's most common deciduous trees is é:yth'elhp, or red alder (*A. rubra*). *Emily Gauthier*

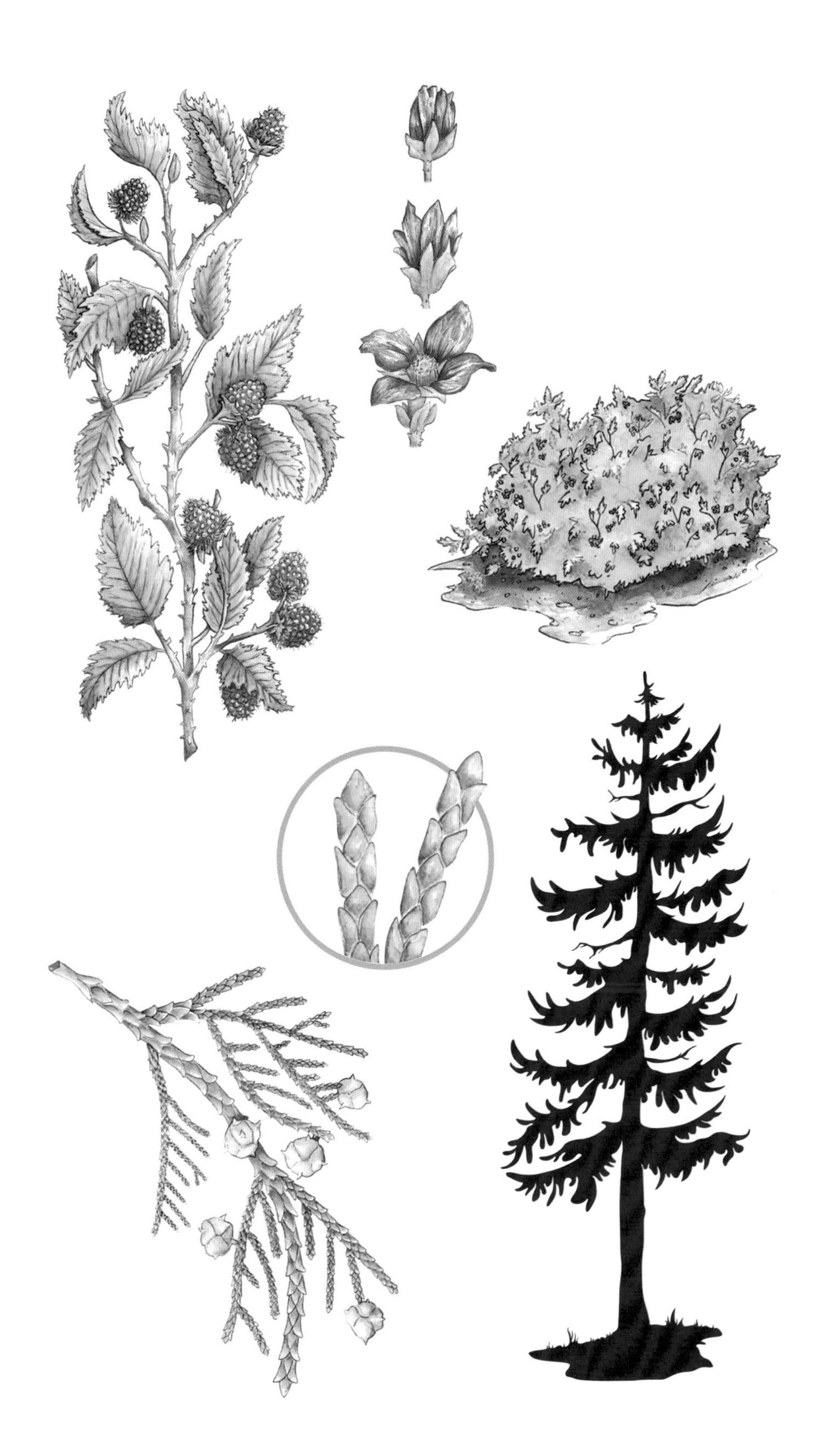

By early March, the season's first arrivals appear, including the skunk cabbage (ch'ókw'e) with its showy lantern-shaped yellow blooms. Fiddleheads slowly unfurl into new bracken ferns. The trailing yellow violet throws out petite sparks of colour close to the trail. The showier magenta flowers of the salmonberry bush (elílá:lhp) are followed by the deeper blues and reds of Oregon grape (th'o:lth'iyulp) and sqe:le, or red huckleberry, in the late spring and early summer.

As summer rolls around again, the trees, shrubs, and mosses of the trail need to reach deep into the understory to draw on reserves of moisture. Crimson-hued net-winged beetles can be spotted feeding on the plant juices of low shrubs, while spiders string translucent webbing between tree branches astride the trail. Hikers swat ahead indiscriminately to clear the nearly invisible spiderwebs.

Sxíxets'—the forest—will experience far less rainfall in the summer. Less than 10 percent of the year's precipitation falls in July and August combined. Short rains make the woods muggy but not muddy. The moisture does not wick quickly from the air and the earth. The summer humidity forms a second skin. What falls is not enough to stop blackberries from shrivelling and blackening on the vine. Mosses pale and then turn brown. Life in the forest slows down and waits until the rainy season begins again.

FACING TOP The arrival of spring is marked by the soft red flowers of the salmonberry bush (*R. spectabilis*), or elílá:lhp. *Emily Gauthier*

FACING BOTTOM Stó:lō communities took great care in using pa:xuluqw, or yellow cedar (*C. nootkantensis*), for clothing and building materials. *Emily Gauthier*

HALQ'EMÉYLEM	COMMON NAME (English)	LATIN NAME
sxíxets'	forest	N/A
é:yth'elhp	red alder	*Alnus rubra*
mólmélhp	western hemlock	*Tsuga heterophylla*
pá:yelhp	western red cedar	*Thuja plicata*
pa:xuluqw	yellow cedar	*Chamaecyparis nootkatensis*
lá:yelhp	Douglas fir	*Pseudotsuga menziesii*
q'emōwlhp	bigleaf maple	*Acer macrophyllum*
slháwél	sword fern	*Polystichum munitum*
st'uslóye	licorice fern	*Polypodium glycyrrhiza*
ptá:kwem	bracken fern	*Pteridium aquilinum*
th'o:lth'iyulp	Oregon grape (bush)	*Mahonia nervosa*
skw'õlmexw	Pacific trailing blackberry	*Rubus ursinus*
qá:lq	rose	varied
ch'ókw'e	skunk cabbage	*Lysichiton americanus*
sqe:le:lhp	red huckleberry (bush)	*Vaccinium parvifolium*
sqe:le	red huckleberry (berry)	N/A
elílá:lhp	salmonberry bush	*Rubus spectabilis*
t'ege'lhp	salal bush	*Gaultheria shallon*
t'áqe	salal berry	N/A

Brent Douglas Galloway, *Halq'eméylem-English Stolo Dictionary*, University of the Fraser Valley (ufv.ca); Brian Compton and Donna Gerdis, *Native Peoples, Plants, and Animals: Halkomelem*, Simon Fraser University (sfu.ca). When there was a discrepancy in spelling, the *Halq'eméylem-English Stolo Dictionary* was used. See also Brent Douglas Galloway, *Dictionary of Upriver Halkomelem* (University of California Publications, Linguistics, v. 141, 2009).

FACING TOP By midsummer, the fruit of the sqe:le:lhp, or red huckleberry bush (*V. parvifolium*), appear. *Emily Gauthier*

FACING BOTTOM The majority of trees planted on the tree farm until the 1990s were Douglas fir (*P. menziesii*), or lá:yelhp in Halq'améylem. *Emily Gauthier*

Habitat

BRITISH COLUMBIA IS home to an abundance of bats, more types of bats, in fact, than anywhere else in Canada. Bats play a critical role in ecosystem health, especially in managed forests like the Mission Tree Farm. Bats eat a lot of insects, including quite a few that most people would rather avoid. Different species of bats feed on moths, mosquitoes, beetles, and an assortment of flies and leafhoppers. In turn, they convert these insects into highly nutritious guano that fertilizes the forest floor.

In the lower Stave Lake drainage, eleven species of bats are found colonizing rock crevices and purpose-built roosting structures. Some types of bats are quite common, including the big brown, hoary, and silver-haired bats. Others are at risk due to habitat loss and *Pseudogymnoascus destructans*, a fungus that results in white nose syndrome, which has wiped out millions of bats across North America.

Despite its proximity to heavily populated areas, the Mission Tree Farm is home to many smaller mammals (bats included), amphibians, and fish. Birds are especially numerous and diverse, and birdwatchers can espy several types of woodpeckers, at least a half-dozen species of warblers, plus bald eagles, red-tailed and sharp-skinned hawks, grouse, jays, thrushes, kinglets, ospreys, herons, owls, ducks, geese, and sapsuckers. Though fewer in number, large mammals roam these woods, including apex predators (those at the top of the food chain), like black bears, and grazers, like the Pacific black-tailed deer.

FACING Forestry staff frequently leave habitat trees like this one standing, as they provide niche habitats for multiple bird species. *Kelly Cameron*

Wildlife on the tree farm relies on riparian areas for feeding. **Wetlands and especially shorelines can also be significant archaeological sites.**

These species are well adapted to life in coastal western hemlock and Douglas fir forests. Some creatures are found in very particular locations on the tree farm and do not travel far over their lifetimes. This is the case for shrews, moles, garter snakes, and red-legged frogs. Cougars, black bears, and other carnivores move between winter and summer habitats and often across tree farm boundaries into neighbouring forest tenures and provincial parklands. Some species show little fear of humans, but more often, critters—such as the western alligator lizard—make themselves scarce around people. The alligator lizard is a small reptile that, from above, shares body proportions but little else with the common alligator. (It only grows to about twenty centimetres in length.)

A century of logging, urbanization, hunting, hydroelectric development, road building, and recreational activities in the region has reduced the size and quality of habitat in the Stave Valley for large carnivores and many birds of prey. Grizzly bears, for instance, were extirpated from the region decades ago and are relatively rare even in Garibaldi Provincial Park to the north. Another species that needs a lot of space but struggles to find it is the northern goshawk, which nests and hunts in mature and old-growth forest. Research studies have been commissioned to assess whether goshawks, marbled murrelets, peregrine falcons, and band-tailed pigeons can be supported on the Mission Tree Farm or surrounding forest lands.

Roads—or, more specifically, the traffic that travels on them—fragment habitat, pose a hazard to crossing wildlife, generate noise and air pollution, and kick up dust. The road network on the Mission Tree Farm is quite extensive, but most secondary roads are gated as a means of controlling vehicle access. As problematic as a busy road can be for wildlife, newer roads are at least designed in such a way as to reduce possible impact on sensitive species. When trucks, ATVs, and motorbikes venture off a maintained road, the effects on wildlife can be significant, especially in riparian zones.

A riparian zone marks a transition between water and land, as is the case with marshes, bogs, shorelines, riverbanks, and intermittent streams. Riparian areas are the most biodiverse habitats on the tree farm. Great blue herons, tailed frogs, painted turtles, and two different species of salamanders can be found along shorelines, in slow-moving creeks, and in cool, damp pockets of the forest floor. Riparian habitats are common in Mission's forest, even at higher elevations, where snowmelt-fed ponds make for valuable watering holes and nesting spots.

Most wildlife on the tree farm relies on riparian areas for feeding, whether on plants, fish, or small mammals, as well as for water and cooling. People, too, are drawn to many of the same spaces, often to collect plants and fish. Because of this, wetlands and especially shorelines can also be significant archaeological sites, as is the case on the west side of Stave Lake. Most people visiting these areas today do so for recreation, although they can unintentionally disturb the sensitive habitats and change the ways that animals behave in a place.

Improperly stored food and garbage put wildlife at risk. Black bear encounters are growing more frequent as more people spend their days and camp overnight on the tree farm and do not always keep their coolers and potato chip bags out of reach. The number of calls to the Conservation Officer Service about so-called nuisance bears spiked in the summer of 2019, after the opening of new campsites west of Stave Lake. They peaked again the following year, after physical distancing and long-distance travel restrictions related to the COVID-19 pandemic drove many into the forests to camp and hike. The waste problem became so bad that overnight use on parts of the tree farm had been suspended by mid-August in summer 2020. Black bears linger near the town landfill, Minnie's Pit, which is adjacent to the tree farm, and nose through trash dumped along logging roads or in surrounding neighbourhoods.

Minnie's Pit also attracts crows, ravens, seagulls, and bald eagles, which can be seen circling overhead at all times of the year. They compete loudly and sometimes violently with one another to lay siege to the rotting food below. Immediately to the west, along Reservoir Trail, ravens have claimed the dominant space high in the canopy—close enough to fly to the landfill and grab a bite. This community of ravens has over the years created one of the most fascinating birding spots in the Mission forest. In addition to guarding the territory against other avian encroachers,

Bald eagles are frequently spotted from all areas of the tree farm, particularly near Hayward Lake and Minnie's Pit. *Kelly Cameron*

the ravens carry on in elaborate and raucous conversations full of cawing, clicking, plopping, and "dwooping" sounds when they detect visitors on the trail below.

Human modifications have most dramatically altered lake habitats on the tree farm. Stave Lake was historically ultra-oligotrophic, meaning that it lacked sufficient nutrients to support a diversity of life. Prior to damming, the Stave River flushed the lake system so frequently that any new build-up of nutrients was limited. Three factors have contributed to better fishing conditions in the Stave today: the construction of the dam, the linking of Stave and Alouette Lakes via a tunnel to control water flows, and fertilizer programs to support fish stocking. Cutthroat, brook, and rainbow trout, as well as pikeminnows and kokanees are found in Stave Lake, and fishing is part of the draw for much of the boat traffic on the water. More patient anglers looking for some peace and quiet will instead set up on the shoreline of Devils Lake and put their lines out for trout.

The dark understory of the forest along the Roy Kittles Trail provides all-season shelter for snowshoe hares and other species. *Kelly Cameron*

Quite a few species of plants and animals found on the tree farm are identified as being at risk or of special concern, under both federal and provincial guidelines. This means that their numbers are considered too low for long-term survival without intervention. Those that are found exclusively or primarily in riparian zones, such as red-tailed frogs or the Pacific water shrew, can be especially vulnerable to logging, trampling, and, increasingly, climate change.

Improving the odds of survival entails investments in habitat protection and enhancement. For species with a large range of movement, providing enough suitable, intact habitat can be difficult, at least in the short term. Arguably, the most dramatic example of this challenge has been the 2007 reintroduction of the Roosevelt elk into the upper Stave watershed. Reintroduction here and in watersheds to the east marked the first time in more than a century that this ecologically and culturally significant animal was found in the Lower Mainland. These efforts to return elk to local mountains, and similar restoration projects, require the resources, skills, and information sharing of multiple levels of government, including First Nations governments. Conservation and habitat improvement efforts are planned for entire watersheds, areas much larger than the tree farm alone.

For day-to-day operations, including logging, Mission works to meet provincial requirements for protecting sensitive and riparian habitats. Overall, Mission's forest is also maturing, which adds to the quality of the habitat for most species. Mission's forestry crews plan timber sales from smaller parcels, to reduce the impacts of harvesting. Afterwards, they build wildlife piles and ponds, using some of the wood debris left behind from logging. It does not take long before wildlife moves back in, despite the absence of forest canopy. Toads colonize these swimming holes and, on a summer day, when the sun bears down and temperatures start to climb, the cacophony they generate rivals that of the ravens at Reservoir Trail.

Adult net-winged beetles flit from bush to bush in search of food, including other insects. *Michelle LeFebvre · all rights reserved*

3 ROOTS

FOR MILLENNIA, COAST Salish Peoples and their ancestors have drawn upon the resources of the Stave Valley and Stave River delta. As the Ice Age came to an end, yellow and western red cedar took root in the till and marine deposits of the lower Stave watershed. People arrived soon after. In the generations that followed, they collected plant resources, cleared the forest by burning, hunted game, and set out traplines.

Despite this long history of settlement and use, no one Nation, village, or family network occupied or controlled the Stave prior to colonization. This was territory that was shared, and at times contested, by many Nations. The Stave was and remains a middle space for Indigenous Peoples

PRECEDING Kwantlen Chief Marilyn Gabriel and Mayor Randy Hawes signing the Memorandum of Understanding, 2015. *Terry Hood*

FACING Along the shoreline of Stave Reservoir is the stump of a cedar tree logged decades earlier. *Terry Hood*

Seventynine Creek in spring. *Jason Brawn*

of the region, visited and settled by families who came from upriver or downriver or from south of the Fraser. The Q'ó:leq' (Whonnock) and sx̱a'ueqs (Skayuk) built year-round and seasonal villages along the river's banks below Stave Falls. The Katzie, Kwantlen, Matsqui, Leq'á:mel, and other Stó:lō Peoples drew sustenance from the Stave and contributed to a dynamic economy in the delta. Despite the absence of fixed boundaries, families collected, cultivated, or managed resources in distinct and often exclusive areas of the forest or shoreline.

Disease arrived before the settlers did, having travelled north and then inland along coastal trading routes by the second half of the eighteenth century. Smallpox and influenza exacted an almost unimaginable toll. After so much loss of life, families and villages merged for cultural and community survival, only to face the other tools of colonization—traders and speculators, forced dislocation and relocation, missionaries, the Indian Act, residential school, railroad and dam builders. Flooding, settlement, and, by the late 1800s, commercial forestry began to destroy much of the visible evidence of the complex Indigenous economy that had once prospered in the Stave.

Gifting

THE CEDAR TREE is called by many names.

Botanists refer to the western red cedar by its Latin name, *Thuja plicata*. Western red cedar enjoys an extensive range; its height and girth are regarded as both awesome and a source of wealth. Because of this, the species has been given many names over the years. These include western cedar, Pacific red cedar, giant arborvitae, giant cedar, and, at one time, shinglewood—this last presumably a reference to the tree's use in making shingles.

In past years, loggers and foresters referred to a second species of cedar, the yellow cedar (*Chymaecyparis nootkatensis*), as cypress, the common name for the ancient family to which the tree belongs—*Cupressaceae*. The tree's territory is adapted to more northerly coastal climes in Alaska and BC, seldom extending south of Oregon. Because of its limited range, the yellow cedar is particularly vulnerable to overlogging and climate change.

In southwestern British Columbia, both species hold significant value to the Stó:lō Peoples. Over thousands of years, language and lifeways shaped distinct chronologies of cedar. The Halq'eméylem language was born here and, by the 1800s, was spoken throughout the central and eastern Fraser Valley. Near the ocean and on the islands and shorelines of the Fraser River delta, the downriver dialect of Halkomelem was spoken. A shared oral tradition tells of the cedar's generosity.

The name for western red cedar tree is pá:yelhp in Halq'eméylem, xhpey'ulhp in Halkomelem; yellow cedar is pa:xuluqw or paxuluqw. The pá:yelhp and pa:xuluqw endowed the Stó:lō and others with a natural source of wealth, and many of the resources needed to adapt to temperate rainforest environments. In a story narrated to Naxaxahlts'i, the late Bertha Peters of Seabird Island near Chilliwack, east of Mission, told of Xepa:y and the origin of pá:yelhp:

> You know that a long time ago there was a very generous man [Xepa:y] who was always giving and always helping people. And they say when he passed away he was transformed into the cedar tree. And because he was such a generous man, that's why we get all the different things from the cedar tree.

Xepa:y's generosity and the bounty of pá:yelhp and pa:xuluqw provided the Stó:lō with the essential raw materials to thrive in the Fraser River basin and surrounding forests. Before any collection occurred, Stó:lō harvesters gave thanks through prayer to Xepa:y's spirit, or shxweli. They used cedar wood for housing, canoes, paddles, and utensils. Women wove stripped roots and split bark into clothing, basketry, rope, fishtraps, and twine. Slewíy, the softer inner bark of pá:yelhp, was prized for use as diapers.

The names given to cedar tell us about how cedar came to be known to the people who use it. For the Stó:lō, Xepa:y's legacy represents an endless gift that requires reciprocity through prayer and gratitude. The European, Canadian, and American settlers in this region lacked generational histories and ecological knowledge. Instead, during colonization, settlement, and harvest, people devised common names based on size, colour, and utility. Scientists assigned other names based on Latin taxonomic classification. These names do not reflect a history of a place but rather categorical naming conventions that consign the trees to their place within a biological hierarchy: Species. Genus. Family. Order.

The cedar tree has many names. The cedar tree is a part of many families. The cedar tree provides order to life in this part of the world.

FACING On the left, a seedling of pa:xuluqw, or yellow cedar. The tree is less common in the Stave Valley than pá:yelhp, or western red cedar (right). *Kelly Cameron*

Bark

THE GIFTS OF the pá:yelhp (western red cedar) and pa:xuluqw (yellow cedar) allowed the Stó:lō to build thriving industrial and trading centres at the mouths of the Stave and Pitt Rivers. In the summer, complex fish harvesting and drying systems ensured a surplus most years of dried salmon. Women and men spent the winter months sheltered in plank houses, manufacturing textiles, baskets, rope, mats, artwork, and boxes constructed for utilitarian and ceremonial purposes.

Not everything woven, built, carved, decorated, shaped, polished, or dried was consumed within the community. The Stó:lō traded widely, with groups living in the Interior of the province and in villages along BC's Northwest Coast. When the Hudson Bay Company arrived, the Stó:lō traded with them too.

Central to much of the production of goods was the sustainable collection of cedar bark (sókw'em) and planks (pó:ys) during the spring and summer months. Bark removal required generational knowledge on how to strip part of the tree without leaving it vulnerable to insect infestation. Before a single cut was made, harvesters performed a ceremony, gave tribute, and often sang songs to thank the tree for its generosity.

Women were traditionally responsible for harvesting, and they selected younger trees for bark stripping. Collecting bark begins with carving notches into the tree with hand tools. Starting at the bottom and moving upwards, harvesters peel the bark away from the tree in long, unbroken

When bark is stripped according to traditional practices, the cedar will continue to thrive for many years while bearing the scar of bark removal. *Kelly Cameron*

strips. Harvesters collect bark from only one side at a time, and in strips no wider than the width of one's hands laid side-by-side on the tree.

Each strip of bark may measure several metres long, and the outer bark is removed by cracking and pulling it away in pieces from the inner bark. The inner bark is then softened through scraping, pounding, and soaking, and rolled and dried for later use. Cutting the strip into narrower bands allows the cedar bark to then be resoaked and woven into the desired end product.

Once the inner bark has been removed, the much lighter bare wood underneath is exposed. Years after these stripped areas have weathered and turned grey, the scars left from stripping are still visible. The impact of cedar stripping can also be detected in the tree rings visible in cut trees. A cross-section from a culturally modified cedar assumes an embryonic shape, as new growth and bark formed around the unaltered sections of the tree. The previously stripped section, meanwhile, stopped growing.

Most cedar bark harvesting today takes place for cultural rather than commercial purposes. Bark is woven into regalia used in ceremonies, for instance, and to educate youth on traditional practices such as sts'éqw' (fine cedar weaving) and the different uses of the tree. The prevalence of CMTs, or culturally modified trees, throughout the Stave watershed and on the Mission Tree Farm suggests a time in the not-too-distant past when harvesting was more frequent and widespread.

Among the most easily identifiable CMTs in the forest are those found near Devils Lake. Archaeological assessment commissioned by the Kwantlen Nation prior to harvesting or facility development has also helped to identify older stripped trees. The oldest CMTs are protected under BC's Heritage Conservation Act, but knowledge of CMTs of all ages is integrated into planning processes on the Mission Tree Farm.

Families

BIOLOGISTS AND VOLUNTEERS with the Fraser Valley Watersheds Coalition (FVWC) spent part of 2016 planting 864 bulbs of sqewth into the re-engineered wetlands in the Stave River delta. For generations, the Stó:lō have cultivated sqewth, or wapato (*Sagittaria latifolia*). This starchy, edible potato-like plant complements a traditional diet of salmon, eulachon, summer berries, game, and other planted and collected vegetables.

Logging, lumber mills, rail and roads, and the construction of two dams along the Stave in the early twentieth century radically altered the wetlands system in which the sqewth grew. Silt deposition, pollution, and controlled water flows changed the delta's ecology. These impacts exacerbated a century-old pattern of loss, displacement, and hunger among the Indigenous families and villages that once managed the resources of the Stave River.

The Kwantlen First Nation worked with the FVWC in planning restoration efforts in the Stave wetlands. The wetlands were reconstructed to provide salmon habitat, re-establish native vegetation, and ensure greater likelihood of survival for the western painted turtle and western screech owl. In addition to sqewth, volunteers planted th'exwíyel (red osier dogwood), willow, qel'qulhp (Nootka rose, or *Rosa nutkana*), sweet gale (*Myrica gale*), and various sedges and grasses. These species add to the biodiversity and habitat stability.

The wetlands were reconstructed to provide salmon habitat, re-establish native vegetation, and ensure greater likelihood of survival for the western painted turtle and western screech owl.

The Stave Valley has been settled continuously since the end of the last Ice Age. Radiocarbon dating provides evidence for settlement of ancestral peoples on the southern end of Stave Lake and along the Stave River, thousands of years before camps developed near the Fraser. Some of the richest evidence has been uncovered from the west side of Stave Lake and the Stave Reservoir, areas inundated by the dam that are accessed today at low-water mark, and in the adjacent woodlands. Artifacts left over from settlements and campfires have been preserved within areas where the forest took root in buried sandbars, dunes, and beaches.

The Coast Salish and their ancestors shaped tools from stone and other durable natural materials, projectile points and knives, abraders, grinding instruments, and body decorations like labrets. Forest environments underwent modification, sometimes on quite a large scale, due to hunting, burning, planting, pruning, clearing, mulching, and harvesting. While most forest fires prior to colonization were started by lightning, Stó:lō hunters and harvesters also set fires to encourage berry growth, enrich the soils, and clear out competing and less useful plants like hemlock and

FACING The Coast Salish Peoples used small-scale burning to encourage growth of important food sources, like huckleberry. *Michelle Rhodes*

heather. Burnt landscapes were easier to navigate for travel and hunting. Hunters employed pack dogs to drive game, and dogs were also used to transport salmon and other goods.

Families and villages held ownership of established plots within low-elevation marshes and estuaries for the cultivation of tubers, rushes, lilies, and beneficial flowers. These wetlands were actively managed and manipulated, and planters dug stream channels to direct water to where it was most needed. These food plots, as well as control over fishing grounds, were inextricably tied up with lineage, power, and influence within and between communities. Often an entire community or settlement owned the resource or land. The Katzie, for instance, collectively managed some but not all of their bogs.

Most cultivated lands were community settlements throughout nearly the full year. Roots and rushes were collected in spring, providing a food source in that critical time before berries ripened. Other roots were stored for consumption later in the year, processed into medicines, or used in trade.

Cedar met hundreds of the day-to-day needs. The Stó:lō sparingly felled large cedars into the rivers and floated them to the Stave delta, where they were used in canoe building. Cedars were carved into totems and cut for building materials, as well as storage boxes and tools. People harvested wood for pit and plank houses, summer camps, and firewood. Women separated the inner bark from the outer bark for use in woven goods, such as clothing, baskets, and rope. Bark and boards were carefully removed so as to ensure a cedar's survival.

The forest provided a bounty of other resources as well. Douglas fir, or lá:yelhp, was highly prized for its strength and durability. The Stó:lō fashioned fir branches into harpoons and hooks for fishing, and they used pitch to caulk canoes. Boiled hemlock bark was used medicinally, employed as a dyeing agent for clothes, and applied to the face. Stó:lō hunters and harvesters established overland collection routes to access a great number of shrubs and fruits, including t'áqe (salal berries), ptá:kwem (bracken fern), moss, and red and black huckleberries.

FACING A stand of maturing lá:yelhp, or Douglas fir (*P. menziesii*), a tree highly prized for its strength and durability by the Stó:lō. *Karin Jager*

While cedar housed and clothed the Stó:lō, salmon fed them. Salmon was sustenance, and fishing was spiritual. The first salmon ceremony, which is still followed today, requires careful handling by selected members of the community of the first salmon caught in the year. After its meat is cooked and shared among all members, the bones are returned to the river so as to ensure the salmon would return the next year.

Sthéqi (sockeye) was most prized as it was the fattest, but all types of salmon were netted. Families erected weirs, fences, and netting systems and smoked the fish to preserve it for consumption throughout the year. Stó:lō fishermen sought out other species as well and told stories of skwō:wech, or sturgeon, that grew "as long as canoes." Swíwe (eulachans or oolichans) provided both grease and meat. In the Stave Valley, fishing was limited to areas downriver from Stave Falls—the falls created too unforgiving an obstacle to overcome. Because of this, the delta was of great importance as a place for fishing, cultivation, and trade.

During spring and summer, small family camps moved higher up in the valleys. There, they gathered the greatest variety of raw materials needed to produce tools, household items, clothing, and ceremonial goods over the winter months.

Stó:lō settlement was about building and thriving in a *system* of familiar and familial places. Year-round survival and economic production required rotating between forest and valley camps, cultivated wetlands, fishing weirs, and permanent village sites. Colonization radically disrupted these cycles of seasonal use and forced upon the Stó:lō a geography of rigid boundaries and confined spaces that cut off access to historic food sources. It also paved the way to the age of widespread logging and of hydroelectricity in the Stave Valley.

Dislocation

THE HALQ'EMÉYLEM NAME Sxyòyeqs translates as "all dead." It is a name that the raven ancestors of today's Kwantlen families gave to the lands along Stave Lake, the place to which ravens migrated after the valley had been emptied of its people.

How the ravens arrived in the Stave is a story in its own right. The Katzie Elder Old Pierre recounted it to an ethnologist, Diamond Jenness, in 1936. The story tells of the arrival of the X̱eX̱á:ls, transformers who would punish the lazy and restless among the Kwantlen. The indolent would continue life as ravens that travelled from camp to camp in search of food. Those who wandered became wolves, tasked with assisting young Kwantlen men and women in hunting and weaving.

The X̱eX̱á:ls were continuing the work of Swaneset, the Creator. Swaneset was both a sky-born hero and the original Kwantlen person, and he descended at Pitt Lake. Here, he created the first Katzie people and built the sloughs from Alouette Lake to just north of the Fraser. The village of the Kwantlen would be here, the place where the spongy moss, q'eyts'i (or Katzie), grew.

Swaneset called upon all of his people to make q'eyts'i their winter home. He beckoned others from farther away to come and stay at this gathering place. Before long, however, there were too many for the village grounds to support. In search of land and resources, the Kwantlen migrated west from q'eyts'i, establishing Skaiametl and Qayqayt (qiqéyt) in present-day New Westminster and north Surrey. Others moved east

The Stó:lō abandoned smaller villages for more central places. Many Stó:lō communities merged or took in survivors from other villages, **despite differences in cultural practices and territorial identity.**

to Whonnock and to the mouth of the Stave, referred to as sxayə'qs, or "everyone's landing place."

In late 1824, scouts working to locate a suitable site for a Hudson Bay trading post found a small village of twenty-two residents along the lower Stave. Just thirty years earlier, despite its abundant riches and its strategic value as a place to build canoes and dry fish, sxayə'qs had been deserted. So, too, were summer camps along Stave Lake and elsewhere, along the Whonnock River and south of the Fraser.

The culprit here, as it was elsewhere across the Americas, was disease. The ravens transformed by X̱ex̱á:ls had returned to an ancestral valley that had been decimated by smallpox.

• • •

By 1782, smallpox and other European diseases claimed the lives of between half and three-quarters of the Stó:lō. Entire villages were abandoned. A decade after the epidemic began, the explorer George Vancouver noted that the landscape had once been home to many more people, although the captain did not know the cause of the depopulation. Survivors of the outbreak carried the marks of the disease. In 1808, Simon Fraser wrote in his journals that many among the Nlhakapmx near Lytton bore smallpox scars.

The area stretching along the Fraser's north shore, between the Pitt and Stave Rivers, was hard hit. Old Pierre told of an "overcrowded" village at the mouth of the Pitt River where disease claimed the lives of almost all the people who lived there. The Katzie Elder told the ethnographer, "If you dig today on the site of any of the old villages, you will uncover countless bones." Parents, children, and Elders, sequestered from others in the community, fell ill and died one by one within their homes.

Old Pierre's son, Simon, shared the location of some of these depopulated villages with anthropologist Wayne Suttles in 1952. These included Snakwaya at Derby (west of Fort Langley on the south side of the Fraser), the Q'ó:leq' community at the mouth of the Whonnock River, the original sx̱a'ueqs community at the mouth of the Stave, and the X̱at'seq settlement on the east side of Mission. Stories of loss and displacement followed the salmon upriver to present-day Chilliwack, Hope, and Yale. One community near Hope was later renamed Sxwoxwiymelh, or "a lot of people died at once."

The Stó:lō abandoned smaller villages for more central places. Many Stó:lō communities merged or took in survivors from other villages, despite differences in cultural practices and territorial identity. People originally from settlements along the Alouette, Nicomekl, and Whonnock Rivers found themselves joining with the Katzie and Kwantlen.

By the 1820s, hundreds of people lived in the Kwantlen villages near present-day New Westminster. They migrated upriver and along the tributaries to collect resources, trade, and create new winter camps. In the summer, hundreds of Cowichan and Squamish canoes passed by the Kwantlen villages, venturing inland to the Pitt River and up the Fraser, where they, too, camped and harvested salmon.

After the arrival of the Hudson's Bay Company near Fort Langley, Tsi-ta-sil-ten, the hereditary Kwantlen leader, and the people of Skaiametl and Qayqayt migrated east. The Kwantlen settled closer to sna'kʷaya, the fort at Derby Reach on lands that had previously been the territory of the Snokomish People prior to the epidemic. The Kwantlen later moved onto McMillan Island, across the river from the fort; the island remains the administrative centre of Kwantlen territory today.

The establishment of Fort Langley drew settlers of many stripes to the Fraser River. Colonists farmed along the river and upland areas. Property was privatized and sold through a system of fee-simple transfers that bore little comparison to land use and allocation systems practised by the Stó:lō. Gold in BC's Cariboo region lured prospectors to the area. Some stayed to take advantage of the lucrative opportunities in wholesaling goods and offering services—including those of ill repute—to fellow travellers. French Oblates, seeking to "save" the Stó:lō from vice and sin, set out from New Westminster to establish a presence on the north bank of the Fraser in 1861, near the present-day townsite of Mission.

The reserve system imposed forced sedentarism upon the Kwantlen and the neighbouring Katzie, Matsqui, Sumas, and Leq'á:mel, starting in 1858.

Before disease and forced relocation, the abundant resources of the Stave River delta supported Sto:lo villages and trade. *Jason Brawn*

The government denied Indigenous Peoples access to the lands previously used for summer camps and resources. The reserves represented a small fraction of the traditional land base. Kwantlen lands were fragmented, with reserves allotted at Albion (353 acres, est. 1864), at McMillan Island (prior to 1862), at Royal Kwantlen Park in Surrey (1878), at qiqéyt in New Westminster (1860s), at Whonnock (1862), along the west side and banks of the Stave River, and on an island in between. No reserve land was allocated along the Stave Lake, in what would become the Mission Tree Farm. No treaty on these lands has ever been signed.

The story of the raven's arrival at Sxyòyeqs tells of a time after smallpox ravaged Kwantlen communities, but it also speaks to the idea of re-establishing a presence in the Stave. The ravens flew up the valley, in search of Kwantlen camps where they were to find food. The Kwantlen and other Nations once planned for a time when they could restore a system of seasonal migration that had fed generations before them. A century of dam building, widespread logging, and road and rails made this vision seemingly impossible to achieve.

4

TRANSFORMING THE STAVE

THE DEMAND FOR the resources of the Stave Valley proved insatiable in the early twentieth century. Early industrial activity in the Stave Valley took two forms: power development and logging. The valley provided an ideal site for electric power generation, and Mission-area forests supplied timber to regional mills. Vancouver's needs required that the space and resources of the Stave be integrated into the metropolitan economy. By the early 1940s, Mission's leaders recognized that a half-century of widespread extraction in the Stave watershed sustained neither trees nor jobs over the long run. A local forest ranger, E.T. Calvert, was among the first to suggest that the community operate its own forest enterprise on lands within the municipal boundaries. The seeds of the tree farm idea had been planted.

PRECEDING View from Mt. Crickmer. *Kelly Cameron*

FACING Transmission towers near Hoover Lake Road carry power east from Stave Falls Dam. *Savannah Powell · all rights reserved*

Power

IN THE LATE 1800s, aside from a handful of cedar harvesters and the occasional prospector travelling upriver to pan for gold at Seventynine Creek, few settlers ventured as far north as Stave Lake. Travelling north from the Fraser, the land gains elevation quickly. River navigation was difficult and portaging around Stave Falls arduous.

To the City of Vancouver, however, the valley was perfect for power generation. Imagine standing on the precipice overlooking Stave Falls, witnessing the Stave River plummet over the twenty-four-metre granite cliff—as tall as an eight-storey building. Imagine how a westerly breeze spreads a light spray from the falls onto surrounding rocks. In all, the Stave River drops eighty metres between the foot of the lake and the entrance to the Fraser. It takes some mental calisthenics to calculate the total energy that could be generated by a dam at this location. The weight of water combined with the force of gravity would come to churn massive turbines.

Vancouver's appetite for power was insatiable. The booming, electrified city needed cheap power to keep up with growth and to enable it to become a hub for industry to compete with Seattle. Companies were anxious to invest in new energy projects. This was generation of power in every sense of the word. Vancouver's growth meant exercising control over rivers well outside city limits.

The Stave was one of several rivers that surveyors, speculators, and engineers looked to as potential dam sites east of Vancouver. Dams were erected on the Coquitlam and Alouette Rivers. The Stave was powerful enough to support two dams. And then, as quickly as it began, the rush for power on the lower Fraser and its tributaries would come to a close with the completion of the Ruskin Dam in 1930.

A small outfit incorporated in 1893 under the name British Columbia Timber Lands and Electric Company purchased land near the falls, with intent to develop the site. Nothing came of this venture, however, and no records exist of the company having even obtained water rights. Two years later, the Stave River Electric and Power Company incorporated with the goal of damming the river, but it encountered political opposition from Mission and Vancouver business interests.

A third group of businessmen formed under the name Stave Lake Power Company (SLPC) in 1899 and gained water rights on the river. Interest in the project ran high, but funding proved difficult to court and sustain. Meanwhile, the SLPC laid out cash for surveys, access roads, bridges, and the construction of a permanent camp for its operations.

The Western Canadian Power Company (WCPC) eventually bought out the entire operation in 1909, and the company operated with the sole purpose of providing power for the growth of factories in Vancouver and vicinity. The WCPC's deep Montreal pockets financed the completion of the Stave Falls Dam project. In its first phase, between 1909 and 1911, dams blocking each of the lake's three natural outlets, an intake dam, and the powerhouse and generating station were constructed. Turbines were spinning by December 1911. Several more years passed before the project was completed.

The first power was delivered to Vancouver on January 1, 1912, via a double-transmission power line erected while the dam was nearing completion. A second power line was constructed from the Stave facilities south across the Fraser River and down to Huntington just north of the US border, in what is today the eastern part of Abbotsford. The Stave Falls generating station did not produce for the town of Mission, however, nor for the rural communities closest to the dam. Steelhead, the rural community uphill and five kilometres east, would not get electric power for another thirty-seven years.

Kwantlen artist Brandon Gabriel tells the story of his people's history and honours the Stave River and its salmon in six panels on Ruskin Dam. *Jason Brawn*

In 1920, the British Columbia Electric Railway (BCER) company bought out the WCPC. The BCER, in addition to operating as a utility, had built and operated Vancouver's electricity-hungry streetcar and interurban rail lines. It emerged as the primary player in the BC power market.

BCER invested heavily in the Stave project, retrofitting the powerhouse in 1924 to add turbines and increase horsepower nearly 250 percent. To accomplish this target, engineers had to impound more water. They did so first by raising the main and intake dams by an additional 7.6 metres. Second, they constructed a dam at Blind Slough just to the east of the powerhouse. Third, and perhaps most ambitiously, they bored through a mountain to create a thousand-metre-long tunnel between Stave and Alouette Lakes. This diverted flow from the Alouette system into the Stave when needed.

To keep pace with Vancouver's anticipated growth, the BCER needed to double production by the mid-1930s and sought out additional dam sites in southwestern BC. The BCER quickly erected the Ruskin Dam on the lower Stave. Shovels were in the ground by 1928, and the turbines were brought on line in 1930. The whole facility—dam and powerhouse—cost only six million dollars and was finished in less than a year.

The dam created a new lake behind it, now called the Hayward Reservoir. Upriver, the raising of the Stave Falls Dam in the 1920s increased the level of Stave Lake and the capacity of the reservoir by an additional 370 million cubic metres of water. The result was a much bigger and wider water body, part lake, part reservoir. Between the Alouette, Stave, and Hayward, the BCER had created a water containment and power-generating system capable of providing a steady supply of energy to Vancouver year-round.

• • •

The Stave Falls and Ruskin Dams replaced the natural churning of water and sediments in the river with regulated releases from massive concrete plugs. Behind the dam, rising waters submerged low-lying forest and erased from view the river's course and the historic shorelines of the much smaller Stave Lake. The Stave Falls and Ruskin Dams were built in a time before environmental assessments. No rules were in place around protection of Indigenous sites within watersheds. The few protections for fisheries were quickly dismissed as being a non-factor, because salmon could not travel on their own up past Stave Falls.

If proposed today, neither the Stave nor the Ruskin Dams would be built. For one thing, the impact on salmon spawning grounds in the Stave delta would likely stop either or both dams. As a younger glacial lake, Stave Lake was likely ultra-oligotrophic—meaning it supported little biodiversity—but the Hayward Reservoir impounded water from what was once a far more ecologically productive stretch of the river. During the 1940s, Mission's Rod and Gun Club repeatedly advocated for fish ladders to be constructed at Ruskin Dam, to no avail. Both the Stave and Hayward Reservoirs today are classed as ultra-oligotrophic.

Between 2012 and 2018, BC Hydro laid out approximately $738 billion dollars to upgrade the Ruskin Dam and powerhouse, reflecting the province's commitment to produce power from the river for several generations more. The Kwantlen and Katzie First Nations have successfully pushed for greater say in the heritage inventory, planning, and management of the Stave and Alouette Rivers. As part of the upgrades to Ruskin Dam, a series of six panels, created by Kwantlen artist Brandon Gabriel, was permanently installed on the dam's face. Each panel tells part of the story of the Stave River and the Kwantlen's relationship to it.

Mission staff continue to work with BC Hydro and the Kwantlen First Nation on areas of common concern along the Stave, including reservoir access, recreation plans, and archaeological inventory. BC Hydro periodically draws down lake levels for archaeological access and to improve fisheries health. Regional non-profits, the Kwantlen First Nation, and government partners have worked to partially restore salmon habitat through the construction of new channels and spawning grounds in the lower Stave. The transformation of the Stave remains an incomplete process, but more emphasis today is placed on planning for values other than hydroelectricity.

Vancouver's appetite for power was insatiable.

The booming, electrified city needed cheap power to keep up with growth and to enable it to become a hub for industry.

In the early twentieth century, loggers carved deep notches into the base of this cedar and inserted boards on which they stood to cut down the tree. *Michelle Rhodes*

Springboards

THE FIRST LOGGERS felled mammoth cedars and towering Douglas firs in the Mission area in the early 1880s. They arrived by steamboat and the recently completed Canadian Pacific Railway (CPR) line. They were Canadian, American, and Scottish, experienced in the forests. They came to cut trees for one of the three New Westminster–based mills that owned or leased land between the Pitt and the Stave Rivers.

In Vancouver, the timber industry was already well established. After adopting the latest milling technologies, BC's first generation of sawmill owners processed increasing volumes of timber coming off Vancouver Island and the south coast. The deforestation of lands along the south coast pushed the industry eastward towards Port Moody, New Westminster, and in the eastern Fraser Valley near Harrison Mills. These mills were followed by others in Surrey and Ruskin at the mouth of the Stave River.

BC encouraged harvesting through an evolving and generous land grant and leasing program. Initially, the province did not charge royalties on timber cut from lands alienated prior to 1887, nor did it prevent the export of unprocessed logs. Licensees paid minimal rent. After 1888, new licence holders paid pennies on the acre and only minimal royalties on cut timber. Railroads were granted high-value, dense stands of Crown forests in valley bottoms and adjacent hillsides. Changes to the size and duration of leases available in 1892, 1903, and 1905 only served to fuel more timber speculation.

At the turn of the twentieth century, a handful of companies dominated BC's timber industry, with major operations in Vancouver, New Westminster, and Moodyville in North Vancouver. One, Fraser River Saw Mills Ltd., acquired the rights to log more than sixty-five thousand acres near the Stave River. An infusion of investment capital transformed this struggling company into Fraser Mills, then the world's largest and most modern milling operation. The Moodyville Sawmill Co., Brunette Sawmills, and others owned or leased land between the Pitt and Stave Rivers. Small, independent mills also popped up in the Whonnock and Haney areas of eastern Maple Ridge by the 1890s.

Early logging in the Stave Valley required experienced loggers capable of manoeuvring oxen and horse teams. At first, crews could harvest only a few of the old-growth giants each day. One cedar might measure upwards of three metres in diameter and up to sixty metres tall. Perched precariously atop springboards, men worked in teams, pulling the long blade of a crosscut saw rhythmically back and forth across the tree's grain. Once felled, a tree's branches and tops were removed, and the log was cut into five-metre lengths. Double-bitted axes and power saws would later speed tree felling.

Logs were harvested close to lakes and rivers, as it was far easier to move the logs by water than over land. When logging crews moved further inland, they relied on a network of plank, or skid, roads. Crews used chokers, chains, and grapplers to hook up logs to animal teams. The oxen or horses then hauled thousands of pounds of dead weight over these roads towards water. To reduce friction created by the drag, workers greased the road timbers with oil or other lubricants. At the shoreline, logs were dumped into the water, boomed, and floated down river to the mills.

To cut more trees more quickly, the combined forces of better sawblades, gas-powered tools, and improvements in rail, steam, and trucking were required. Money flowed into the industry from investors, and with it, a reorganization of timber leaseholders. Widespread harvesting provided more jobs for residents of the nearby rural communities of Steelhead, Stave Falls, Silverdale, and Ruskin.

To cut more trees more quickly, the combined forces of better sawblades, gas-powered tools, and improvements in rail, steam, and trucking were required.

By 1900, the largest timber operator near Mission was E.H. Heaps & Company Ltd. Edward H. Heaps had assumed control of the Lobb Shingle Mill at Ruskin two years earlier. Heaps was new to the timber industry, having worked in England's cotton manufacturing sector prior to setting sail for Canada. When he arrived in Vancouver, Heaps purchased a pair of shingle machines for cedar. From there, he expanded into milling.

Financial woes forced Heaps to sell his timber leases east and west of Stave Lake after 1909. Before then, however, Heaps introduced steam donkeys and rail to the valley. The steam donkey resembles an armless Tin Man cobbled and rivetted together from steel panels and shaped into a tall cylinder with a narrower stove pipe on top. A wood-fed boiler was outfitted at the base to power a spooled winching system at the front of the donkey. The whole apparatus was affixed to a heavy sled that could be towed along the skid roads.

Steam and later gas donkeys allowed crews to haul logs over longer distances. "Donkey punchers," or machine engineers, operated the winches and warning whistles. Workers uncoiled and extended cables great distances through the woods. Specially trained men called choker setters

A rusty nail and vibrantly hued lichens mark the passage of time for this century-old plank road east of Stave Lake. *Justine Robinson · all rights reserved*

attached cables to logs to be hauled. Or they hooked cables to standing trees, and donkey punchers set the winches in motion. Cables could also be run up to the top of an A-frame system. Crews could hoist a log and manoeuvre it onto a pile, down into the river, or onto a train car.

After Heaps's company went under, his timber leases were bought by two local interests. The Keystone Logging Company logged Heaps's former leases in the Silverdale area between 1915 and 1922. The Abernethy Lougheed Logging Co. purchased Heaps's timber leases west of the river. The firm incorporated with the specific goal of harvesting within the Stave and Alouette watersheds.

• • •

The standard-gauge railroad provided a faster and safer alternative to the steam donkey. Between 1906 and 1924, the number of logging railroads built in the Fraser Valley doubled, from six to twelve, including a line Heaps built north from Ruskin. Rail builders drove tall timbers into the mud and gravel of the Stave River to serve as pilings for the rail bridge.

Train cars would haul the logs out to the middle of the bridge, half a mile from the river's mouth, and dump them over the side into the water. The logs were then boomed and towed downriver.

The Western Canadian Power Company built another line in 1910, while building Stave Falls Dam. At just under six miles (ten kilometres), running parallel to the river's shore, the company claimed this as the world's shortest chartered railroad. The WCPC line ran on electric, steam, and gas engines and used skeleton cars for hauling logs; the trains moved mail, people, building supplies, and logs between Ruskin and Stave Falls.

Rail served as the highway along the Stave until the 1940s. The primary beneficiary of the line was the Abernethy Lougheed company. Starting in 1915, the company used the A-frame logging technique to harvest timber along the lakeshore. Abernethy built a sawmill and shingle mill just north of the Stave Falls Dam. At peak, the company moved more than one million board feet (2,360 cubic metres) of cedar and fir by rail each year.

Like the steam donkey before it, the railway would soon be replaced by speedier options. Five-ton trucks were in limited use in BC by 1915, their expansion held in check by the lack of quality roads and restrictions on US vehicle exports during World War I. By 1920, truck logging arrived when W.E. Parsons began transporting timber poles from the back of Mission along a privately built road using a new two-ton National truck.

Each new advance in transportation marked an expansion in the scale and geography of harvesting. Crews worked faster and penetrated deeper into the forest, using winches, chainsaws, and the internal combustion engines. Log booms choked the shorelines of the lower Stave River. Workers became more specialized, in line with the skills and experience needed to operate winches, climb trees so as to rig the cables needed to hoist and move logs, or drive a truck down a steep grade with several tons of weight bearing down on the axles.

The Stave Falls Railway stopped service in 1944, after more than thirty years of operation, marking the end of railroad logging in the valley. The steam and gas donkeys were also retired. A new age of modernized logging had arrived.

Shakes & Shingles

LEFT UNTREATED, CEDAR wood fades from copper to grey and white with age, but it largely retains its structural integrity. Cedar is lightweight and durable, an ideal material for roofing in wet climates. The Stó:lō knew of the cedar's generosity and its utility for building homes. By the early 1900s, settlers and loggers also recognized its potential.

The technology to mechanically transform cedar into shingles—squares of wood only two centimetres thick—had been around since the 1820s. Shingle mills were in operation in Victoria by the 1870s. Shakes, similar to shingles, are thicker and only partially shorn. BC initially exported much of its unprocessed cedar to state-of-the-art high-volume mills in the US. Canadian mill capacity expanded, and by the early twentieth century, most of the cedar harvested near Mission was being turned into shakes and shingles by domestic mills.

The processing of cedar into shakes and shingles began in the forest. F.C. Carter, an Ontario native who migrated to Whonnock in 1889, hauled oxen for E.H. Heaps & Company. He later told of how loggers prepared cedar for transport. After trees were felled, the next steps involved "cutting the logs 16 feet long, splitting in quarters and taking the heart out. A 'long-bolt' could not have more than five knots in it. These logs were 'dogged' up in a line about six long and hauled by oxen or horses for two miles to the Stave river, boomed at Ruskin and towed in boom to New Westminster."

Flume systems from the early twentieth century were still largely intact in the 1960s, though long out of use. Little evidence of them remains today. *Mission Community Archives 0305-1*

Splitting a western red cedar tree into bolts for processing into shakes or shingles.
Mission Community Archives 0305-2

Alternatively, the log would be further subdivided into shorter lengths of a little more than 1.4 metres. The cutters would continue splitting cedar, until each bolt was not more than 1.2 metres around. The bolts were then bundled into cords, or bundles of thirty-two bolts, as if piecing the tree back together again.

The work of a bolt cutter was arduous, repetitive, dangerous, and low paid. The mills and logging companies employed crews of recent arrivals, young men new to forestry, and Japanese and Chinese immigrants. At the end of a day's work in the woods, workers crowded into small shacks, bunkhouses, and camps. Poorly lit and underventilated, camp conditions deteriorated from their already dismal state as the season progressed.

Until truck logging came along, animal teams and steam donkeys hauled cords of cedar out along skid roads. Elevated flumes were also erected to transport cedar bolts. Crews constructed the flumes out of cedar, balsam fir, and hemlock, and lake or creek water was pumped into the flume to flush the bolts downslope.

Decades later, remnants of the flume systems were still visible in the forest. Lorne Rockwell, who worked for the tree farm starting in the late 1960s, recalled coming across a flume along Rockwell Creek. (The creek was named after Lorne's father, Irwin "Rocky" Rockwell, Mission's first tree farm manager.) A few miles up from the Rockwell delta, where the creek empties into Stave Lake, tree farm crews found a shallow, five-acre lake that had been partially dammed.

"At one end of the lake, the Japanese had built a weir out of cedar, split cedar," Lorne Rockwell said. "Now the reason they built this weir up there was to build the lake up. The weir was constructed of all cedar, all split cedar that they hand split... and it would build the lake up about two feet. And what they had was little flat boards... There were lots of creeks feeding that lake up there."

Long cedar boards restricted the flow of water down the creek. When workers were ready to move the cut cedar, "they'd pull the boards and the water would sluice down and take the shingle boards down."

One of the largest flume systems in the Lower Mainland was built by the Stoltze Manufacturing Co. Flumes varied in distance, including one five kilometres long originating at Rolley Lake. Crews moved bolts to the Stoltze Mill north of Ruskin. In his time working as a young man for Stoltze, A.S. Cameron (later a prominent mill operator in Mission) recalled that the cedar was to be used to build storage boxes for batteries, because cedar wood was more resistant to battery acids than alternatives.

Not all shingle mills were as large as Stoltze's, and smaller operations were scattered across north Mission in the 1910s through 1930s. In the decades that followed, cedar mills would locate almost exclusively on the Fraser River. In the forests, cedar bolt cutting and salvage operations provided a bulwark against the worst effects of the Great Depression, when most other parts of the timber industry limped along. The building boom that followed the end of World War II brought renewed vigour to cedar shake and shingle mills, however, and Mission emerged as one of North America's biggest centres for shingle and shake manufacturing.

Mills

IN THE 1980S and 1990s, battles erupted in BC's coastal forests over whether to log the remaining stands of old growth. Environmental activists and workers alike questioned why ships full of raw logs sat in ports awaiting transport to US or overseas mills. The cargo represented the loss of both healthy ecosystems and good-paying jobs.

But this wasn't a new fight. The concern over raw log exports dates back to BC's early years, and for many of the same reasons. The provincial timber industry emerged in the shadow of massive American mill complexes that sprung up in Seattle and Portland starting in the 1860s. US mills could manufacture wood products at a much lower cost than could the newer but smaller Canadian mills. In response, BC eventually prohibited new licensees from exporting raw logs extracted from Crown coastal forests.

Producers struggled to find buyers for processed wood, and the two countries have never enjoyed free trade in manufactured softwood products. American quotas and tariffs were applied to dimension lumber, shingles, and pulp. On the other hand, hungry American mills happily imported as many *raw* logs as BC would send them, as these were exempt from US restrictions. The US federal government worked to prevent competition in processed goods. For instance, in the 1930s and 1940s, Canadian red cedar shingles could account for no more than 30 percent of those consumed in the US marketplace.

Other changes in the timber industry and government policy compounded the difficulties for Mission's mills. Operators faced supply issues,

Lumber mills have operated along this stretch of the Stave River continually since the early 1880s. *Jason Brawn*

frequent labour action, new technologies, and international competition. Changes in tenure and licensing rules for Crown forests favoured large companies over smaller operators like those in the Mission area. Mission's wood products industry seldom enjoyed long-term stability, as mills frequently burnt down or were sold, temporarily shuttered, rebuilt, reorganized, retooled, torn down, and all but erased from the landscape altogether.

• • •

Despite its advantageous location, Mission failed to attracted investment from the major players in BC's forest industry in the early or mid twentieth century. Its industrial waterfront lands did not lure them, nor did the District's proximity to the US border or its location at the junction of north-south and east-west train lines.

Companies like Fraser River Saw Mills, Canadian Forest Products (CanFor), and H.R. MacMillan Export Company established major milling complexes west and east of Mission. At times, some held timber leases in the region but with little or no intention of processing wood locally. Instead, logs were floated down the Fraser to milling complexes at New Westminster, Vancouver, Richmond, and Surrey. Mills established in Mission and the Stave Valley were small and mid-sized operations. The owners usually lived in Mission or Ruskin.

Milling started locally in the 1880s. By 1892, Lobb Shingle Mill operated in socialist leader John Ruskin's eponymous community near the mouth of the Stave. A decade later, the mill was sold to E.H. Heaps & Company.

Across the river, George McTaggart, a Mission resident, built a mill on the Silverdale flats in the 1880s. Decades later, Silverdale resident W.B. Chester, whose father had homesteaded the area, recounted:

> [The McTaggart mill] had a switch on the CPR and also a road down to the Fraser River. The road ran south to the river from my crossing over the track. Between 1894 and 1896 the mill changed hands and was run by Welband and Purdy as the "Sawmill Co. of Silverdale," at least this was the name my father used when he sold them hay, according to his account book. His last entry was May 7, 1896... They got in wrong with the CPR and the CPR wouldn't take their ties as they left a lot of ties piles alongside the tram road when they moved out.

Dozens of additional mills sprang up in Ruskin, Whonnock, Silverdale, Steelhead, and east of town, in Hatzic. Many companies relied on fir and cedar harvested on lands only a short distance from the mills. The Solloway brothers erected a small mill near 7th Avenue in Mission at some point after 1905. In the 1920s, mills specializing in railroad ties and cedar shingles were built by John Cade, the Barr brothers, and Peter Hammer on the city's north side and in Steelhead. Another mill was built along the shore of a small lake at the foot of Bear Mountain, at what would later be called Mill Pond. Later, Townsend "Toke" Meeker built the first cedar shake mill on the flats south of downtown Mission. Few of the sawmills survived past the 1930s.

Nels Anton, a Swedish immigrant, established a pole business in Ruskin, near where W.F. Watkins would later build Watkins Sawmills.

The provincial timber industry emerged in the shadow of massive American mill complexes that sprung up in Seattle and Portland starting in the 1860s.

The Stoltze Mill was built a less than a kilometre north of the old Heaps sawmill. Closed in the teens after the company ran into financial trouble, the Heaps mill sat vacant until purchased by Toke and Henry Meeker in 1935. They named their operation the Stave Lake Cedar Mill, a name they retained even after they relocated to Hatzic in 1939.

Mills also sprang up within the present-day boundaries of Mission's forest tenure, years before the province awarded the municipality Tree Farm Licence 26. The largest of these outfits was the Abernethy Lougheed Mill, just north of the Stave Falls Dam. This site was one of a few mills along the lake's southwestern shore operating between 1915 and 1964. Rising waters behind the dam claimed the first of these, Abernethy Lougheed's, in 1926, although by that point, the company had shifted operations to a much larger, more modern facility at Alouette Lake.

Smaller mills staked out territory along the shoreline on both sides of Stave Lake. The Ebbisuzaki family owned one of these, although it had burned down by 1928. The creosote timbers left behind by one mill remain buried by the gravel at the shoreline of Rockwell delta. At least two other mills were lost to fire. When the last two mills ceased operations at the site in the 1960s, no new facilities were built, as this stretch of shoreline had become part of the tree farm.

Migrations

IMMIGRANTS FROM ASIA and Europe laboured in Mission-area logging camps and mills from the 1880s through to the 1960s. Experienced loggers, mill foremen, and manual workers were all needed. Swedish, Japanese, and Indian migrants developed expertise and became leaders in the industry, and some went on to invest in forest companies of their own. The Mission Tree Farm later benefited from Scottish immigrants trained in European forestry.

Artifacts left behind by Chinese and Japanese workers can still occasionally be found in the forest—a piece of rusted silverware, a chipped porcelain teacup half buried in dirt. Employment in the forests was highly unequal. The experienced crews responsible for planning, cutting, and loading timber were largely white and Canadian or European. The most recent arrivals from Asia were assigned many of the lower-paid, menial, and dangerous jobs in the forest, including cedar bolt cutting. Even when working the same jobs as white workers, Japanese and Chinese workers earned far less.

The area's mills benefited from this immigration as well. In 1967, Stan Carter, son of Whonnock homesteader and oxen hauler F.C. Carter, shared his early memories of the E.H. Heaps & Co. Mill with the *Fraser Valley Record*:

> Just north of the CPR trestle was the Heaps' lumber mill, a mill of some 30 men, mostly Chinese, with a turnout of about 23,000 bf [board feet] per day of 10 hours. Then there was a dry kiln, two Chinese bunk houses and a cook house.

FACING One-room shacks like this one near Steelhead were built by immigrant loggers and miners and once dotted the Mission forest. *Kelly Cameron*

How many Chinese worked in Mission-area mills is not known, but by 1901, more than 2,500 Chinese immigrants were counted in New Westminster and the Fraser Valley. Carter's memories of Ruskin suggest that Chinese workers spent at least some time in the Stave Valley. Most Chinese immigrants eventually moved closer to Vancouver or elsewhere in Canada.

More than two thousand first- and second-generation Japanese Canadians—the *issei* and *nisei*, respectively—made Mission their home during the 1910s through the 1930s. Their first jobs were often in logging camps near Haney and Whonnock in Maple Ridge and in the Stave, Cedar, and Hatzic Valleys of Mission. Most Japanese immigrants to Mission worked in the timber industry at some point, and at times, husbands and wives worked alongside one another in logging camps. More commonly, Japanese families purchased farmland, which women maintained while husbands and sons spent some or most of the year labouring in mills or in the woods.

In the 1920s, *issei*-owned mills began appearing in Maple Ridge and Mission, including in the Stave Valley. Yojiro Takimoto and Kahei Kamimura each established multiple factories producing lumber and railroad ties. Like most sawmills built in Mission during this time, these mills did not survive fire or the Great Depression. Had they succeeded, their owners would have had to sell or been forcibly dispossessed of their business properties as a result of wartime internment orders issued in 1942. Following their detainment and expulsion from coastal zones, few *issei* and *nisei* returned to the Fraser Valley.

Mission's Sikh community similarly traces its Canadian roots to the expansion of sawmills and shingle mills in the 1930s and 1940s. Across the province, migrants from northern India accounted for a large share of the millworkers. Mill owners preferred new immigrants from the Punjab in part because they paid these workers significantly lower wages than their Euro-Canadian counterparts. Many South Asian migrants established roots in smaller BC timber towns like Duncan, Port Alberni, Powell River, and Mission.

Recent overseas arrivals worked ten-hour shifts in the mills, six days a week, and often resided with other South Asian workers in bunkhouses. They sent for wives and children as soon as they could afford to. Many of

Artifacts left behind by
Chinese and Japanese
workers can still occasionally
be found in the forest—
a piece of rusted silverware,
a chipped porcelain teacup.

Herman Braiche erected his mill and barge loading facility along the Mission waterfront, on the land just west of the CPR bridge. *City of Mission*

their Canadian-born children also worked in the mills, and immigrants went on to establish mills and related businesses of their own.

Some South Asian immigrants rose to prominence in Mission's business community. Indar Singh Gill (1903–93), who was only seventeen when he left India, spent nearly two decades working on farms near Kamloops and in sawmills and on Vancouver Island. He moved with his family to Mission after the war. Using savings from his time in the mills, Singh built a company that delivered sawdust to people's homes for home heating. This undertaking evolved into Indar Fuels, and the company's success provided Singh the capital to make bigger and more lucrative investments in sawmilling on the Mission waterfront in 1959–60. Singh also served with Mission's East Indian Community Association. While his Fraser Valley Sawmill was lost to fire in 1964, Singh later started a shake and shingle mill in Fort Langley.

Like Indar Singh Gill, Herman Singh Braich (1911–76) migrated from India as a teenager, landing in BC in the late 1920s. Braich worked in mills on Vancouver Island before moving to Mission after World War II. He purchased land on Mission waterfront after the 1948 floods, on which he built a mill and barge-loading facility. Braich supported new arrivals from India, providing them with work and bunkhouse accommodation. Fire claimed Braich's mill in 1968, and it was never rebuilt. Most of his employees crossed the river to Abbotsford, which had a growing Sikh community and more job opportunities. Braich and his family remained in Mission and stayed involved in community organizations.

Naranjan Grewall (1908–57) migrated from India via England in 1941 and soon established himself as a union-friendly business and political leader. Grewall owned Mission Sawmills Ltd. and several other companies in Mission and Vancouver. Before fire tore through his buildings and log stockpiles in 1947, Grewall's waterfront sawmill operation was supplying half of Mission's wood fuel and 40 percent of the lumber used in local construction. The firm also held a major contract to send finished lumber to Great Britain.

Grewall's mill was rebuilt, only to burn down again four years later. This time, Grewall chose not to rebuild, but he did agree to manage Fraser Valley Pulp & Timber Ltd. when it was established, several years later, at the same location. During this period, his business holdings grew to include Mission Forest Products Ltd., which logged in Whonnock and Harrison, and a shingle and shake mill west of Stave Falls Dam.

Grewall used his business influence to press for political change, including advocating for local forest control. He successfully ran for the Village of Mission board of commissioners, winning the most votes of any candidate in the December 1950 election. Grewall was likely the first elected official of South Asian descent anywhere in Canada. In 1954, he was named board chair. Two years later, he ran unsuccessfully for provincial office.

Grewall, Singh, and Braich invested in the latest technologies and contributed to the modernization of Mission's wood products industry. They built on the Mission Flats, the low-lying, flood-prone lands adjacent to the Fraser River, at a time when many smaller mills were closing or locating in other communities. But when fire claimed their waterfront mills—Grewall, Singh, and Braich had each lost at least one mill this way—rebuilding was not always practical. By the 1960s, Mission struggled to hang on to its wood products industry.

5

THE CASE FOR CONSERVATION

THE 1950S AND 1960s brought a new level of prosperity to British Columbia. Bolstered by postwar industrial expansion and the baby boom, BC's forest industry was enjoying some of its best years on record. Confident about the industry's longevity, workers in timber-dependent communities from Vancouver Island to the Peace Country sank their paycheques into their first homes, new cars, and family vacations. That any timber town would choose to establish its own community forest during this time seems, on the surface, risky if not foolhardy. Pursuing a tree farm licence required investments that had no guarantee of return. In exchange for control over Crown forest lands, the municipality was required to absorb the costs of road building and fire prevention.

PRECEDING Pond on Mt. Crickmer. *Kelly Cameron*

FACING Early 20th century logging outfits transformed the old-growth forests of the Stave Valley. *Savannah Powell · all rights reserved*

Mission's pursuit of a community forest meant putting some timber production under public management, when timber harvesting had up to that point been entrusted almost entirely to the private sector. District leaders were not anticapitalist or even anti–big business, but they questioned the health and sustainability of an industry increasingly dominated by large firms and unsustainable practices. If they were to help make a future possible for Mission's smaller operators, they would have to ensure that there would be enough trees available to cut in the future.

By the late 1960s, the Mission Tree Farm used small-scale clearcutting, which was more efficient and made replanting easier relative to past practices. *Mission Community Archives 0305-3*

Sloan Commission

MISSION'S POLITICAL LEADERS first explored the possibility of establishing a community forest during the later years of World War II. In 1943, the Mission and District Board of Trade invited E.T. Calvert, a seasoned ranger with BC's Department of Forests and a Mission resident, to speak to its membership. Calvert raised concerns about the rapid expansion of the province's forest industry and argued for the benefits of community forests like those seen in Sweden and elsewhere in Europe. Following Calvert's presentation, the board established a working committee on forestry alternatives, staffed with local logging and sawmill interests.

The next year, the board invited a second forester, F.S. McKinnon, to speak at length on community forests. In his speech to the board, McKinnon asked:

> What does a community forest offer to local citizens? These forests will offer opportunities for coordinated and complete utilization of all the forest values and resources. In other words, the forests will be used for profit, pleasure and health. Thus, the objective will be to obtain the maximum of financial return as well as the optimum of mental and physical relaxation which may be associated with forests.

At the time, the owners of small and mid-sized timber companies—businessmen like log shipper A.B. Catherwood—dominated local politics.

BC had for years struggled with the right balance between supporting industry and trying to regulate it and manage timber supplies.

Catherwood was later elected reeve of the Village of Mission. (The Village of Mission is the downtown core and surrounding neighbourhoods of what became the District and later the City of Mission.) He and other operators questioned whether poor forest management would leave their companies and others without a reliable timber supply.

The District decided to convert a few thousand acres of cutover lands, properties it had come to own through tax defaults, into a local forest venture. But it lacked legal authority to do so. Hoping to pressure the province to make needed changes to its Municipal Act, in July 1944, the Mission District's council unanimously passed a measure in favour of establishing a community-managed forest. Mission's leaders also reached out to Fraser Valley political and business leaders to try to cultivate wider support within the region.

Any needed changes to the Municipal Act, however, would have to await the results of a provincial inquiry into forest health. On the last day of 1943, the province called upon the Honourable Gordon McGregor Sloan, BC's chief justice, to lead a commission charged with investigating the state of BC's forests. The Sloan Commission, as it came to be called, was to examine and recommend changes to forest inventory and tenure systems. Sloan and his team had to confront two seemingly contradictory questions. First, how does the province ensure enough wood will be available for growing demands? Second, how should BC contain speculation and reduce the rapid rate of depletion, especially along the coasts?

BC had for years struggled with the right balance between supporting industry and trying to regulate it and manage timber supplies. In 1907, the BC government convened its first royal commission to examine a boom in timber speculation. Recommendations from the resulting Fulton Report, issued in 1910, shaped the 1912 Forest Act, which outlined the rights and obligations of future timber sales licences. The Act slowed but did not stop the depletion of BC's coastal forests, and by the 1940s, companies confronted diminishing timber supplies. Most of the iconic, centuries-old giant cedars and firs were gone. Companies moved uphill and inland and began harvesting immature or lower-value stands.

Forester F.C. McKinnon, in his speech before the Mission and District Board of Trade in 1944, warned board members of as much:

> It is at this point that the odd corners which were not logged the first time over are given careful appraisal and, in most cases, are logged. Also, the stands of second growth which are easily accessible are cut even though they would yield much better if left for another 20 or 30 years.

Up and down the coast, harvesters logged smaller and smaller diameter trees. Huge volumes of wood waste dried out and sometimes caught fire. The Great Depression put many small timber outfits out of business; others adopted a "cut and run" attitude. In Mission, speculators bought land to harvest its high-value timber before abandoning the properties altogether. During and after World War II, the wartime effort and economic expansion magnified the pressures on coastal forests.

In its 1945 report to government, the Sloan Commission issued a call for sustained yield management. In simplest terms, this refers to not cutting more than is planted in order to ensure that forests generate commercial yields in perpetuity. Up to this point, reforestation was rare and often ineffective. For leaseholders, it simply made better business sense to move further inland to harvest new timberlands than it did to replant.

To incentivize the investment needed to make reforestation and sustained yield a success, the Sloan Commission recommended two new types of tenure. The first was the public working circle, in which smaller logging companies operated within a larger sustained yield unit (or "timber supply areas," TSAS) managed by the BC Forest Service (BCFS). The second type was private working circles, in which individual firms could combine existing leases and privately held lands with newly issued tree farm licences to create a more rational and orderly management system. The licences were to be issued for twenty-five years or longer. For both TSAS and TFLS, annual allowable cut limits (AACS) set caps on total harvesting in a given year.

Mission came into ownership of forest parcels like this one in Steelhead after previous owners took valuable timber but then defaulted on their taxes during the 1930s.
*William Van Nieuwkoop · *

By design, the new tenure system favoured the bigger timber operators, who had the deep pockets needed to invest in TFLs. Using this type of licence, the companies soon captured much of the available higher-value timber supply in the southern half of the province. As a result, these long-term leases tied the fates of many resource communities to those of the large timber operators.

In its report, the Sloan Commission expressed support for community forests and proposed that public working circles met this need. The chief justice wrote that "community forests, apart from the timber production therefrom, have proven to be of considerable value in the United States as a means of acquainting the public with the benefits to be secured from the practice of sustained-yield forestry, the necessity of fire protection, and related subjects."

Sloan's tenure reform did eventually create an opening for Mission to develop its own community forest, although not, as expected, through a public working circle licence. The District opted instead to first build the Mission Forest Reserve.

By the 1940s, local mills questioned whether they would be shut out of accessing high-value timber close to home, such as these firs harvested in the late 1960s. *Mission Community Archives 0305-4*

The Forest Reserve

AS EARLY AS 1944, board of trade president A.G. McInnes called for "an inventory of the forest resources of [the] district and a summary of the lands held in public ownership." E.T. Calvert was hired in 1948, following his retirement from the BC Forest Service, to complete the assessment of the state and value of District-owned forest lands. Most of these were in Steelhead and Stave Falls.

Calvert's 1948 inventory of the tax-default lands generated the information needed to create a reserve. But before council could put the measure before voters, the province had to first approve changes to the Municipal Act that would allow for the establishment of municipal forest reserves. In summer 1948, the Union of BC Municipalities (UBCM) added language stipulating that "the Council of any Municipality may by by-law set aside as a forest reserve any lands owned by the Municipality that in the opinion of Council are suitable for reforestation purposes."

With the legal hurdles cleared, Mission's voters headed to the polls on December 18 to vote on the proposed forest reserve. Reeve Catherwood had spent the weeks beforehand talking up the Swedish model. He touted that "Mission Municipality may one day have a similar experience when the Forest Reserve Bylaw brings out municipal-owned forests to full annual production." The sales pitch worked. More than 90 percent of the 672 votes cast were in favour of creating the Mission Forest Reserve using 2,660 acres of tax-default lands.

The forest reserve was just a start. On their own, the municipal lands would never be big enough to provide more than a negligible supply of salvage timber, and certainly not enough to sustain more than a handful of seasonal jobs. Most of the marketable, high-value timber had been logged off reserve lands years before. It would be another forty or fifty years before much of the forest could be harvested again.

Still, the reserve allowed the District to experiment with improving harvesting and replanting practices. Mission soon become a darling of a national forest conservation movement. The Canadian Forestry Association organized train trips for its members to tour the Mission Forest Reserve. The reserve became one of only two dozen such projects to be certified as a Canadian tree farm by the association by 1955. Achieving economical sustainability, however, required expanding onto the Crown lands lying just to the north of the reserve. The District pressured the UBCM to make further changes to the Municipal Act to allow for local control over Crown lands.

After only a few years of running the forest reserve, the District would apply for a tree farm licence. They would do so just as the community's mills were undergoing major changes that threatened their very stability.

The Hemlock Economy

IN THE 1940S and 1950s, Mission's most prominent political voices were also independent operators of logging, milling, and towing firms. Albert McMahon, Herman Singh Braich, and Naranjan Grewall built mills on the industrial flats along the river, where the Catherwood family also ran a successful boom and towing business. McMahon, who served as president of the Mission City and District Board of Trade, was active in the Mission School Forest Association (MSFA), alongside Albert Thomas of Thomas Lumber. (McMahon was also instrumental in Mission's application for a tree farm licence.) Braich helped recent immigrants from India find work in his mills and establish roots in the valley. Both A.B. Catherwood and Grewall served terms as reeve for the Village of Mission.

Some of these same business owners sounded the alarm over the dangers posed by the large forestry firms operating in southwestern BC. Their worry, at least publicly, was not about competition in the marketplace. Instead, they voiced concern that wood supplies would be effectively locked up in long-term leases under the new provincial timber supply management system.

The changes to the tenure system introduced after the Sloan Commission's report provided order to the mosaic of land tenures found in coastal forests. However, these changes put many small mills and logging outfits at a disadvantage in applying for a tree farm licence, because they could not afford to take on a larger tenure.

Other small operators complained that they were shut out of the system by political leaders, as tree farm licences were issued at the discretion of the minister of forests. Mission City and District Board of Trade president McMahon pointed to the problem of speculation by tenure holders who did not actively log. W.F. Watkins, owner of Watkins Sawmills, called for the creation of an appeals board in part because his own application for timber in the Agassiz area had been unsuccessful. His complaint was less that he had been outbid than that the timber had yet to be logged several years after the licence was issued.

Grewall vocally opposed the granting of tree farm licences to large operators, whom he likened to "Timber Maharajahs." When Grewall ran for provincial office, the Dewdney CCF council professed that Grewall "has consistently fought for the rights of the small logging and sawmill operators. He has fought their battles before the Sloan Commission against the ever-growing monopolies caused by the present methods of awarding forest-management licences." (The CCF, Co-operative Commonwealth Federation, was the precursor to the NDP.)

Local mill owners and logging firms pointed to other concerns as well. The BC Truck Loggers Association, which once represented the interests of smaller, independent operators, put timber supply issues front and centre at its annual meetings. Among other factors, the association pointed to overly conservative estimates of timber supply that had limited harvesting potential. Others worried that investments in replanting did not generate sufficient returns within the lifespan of a timber licence.

Changing mill technologies added to supply woes. Mills across BC were increasing production volumes, driving down unit prices. Newer, integrated mill complexes could concurrently manufacture plywood and dimension lumber. McMahon told a community group in Mission in 1951 that at his mill, "it takes 35 to 40 men to produce half-a-million dollars' worth of forest products a year." Other companies invested in new machinery that could run 24-7, requiring fewer workers but more wood. Larger firms maintained a significant cost advantage over smaller and aging firms.

By the mid-1950s, Mission's mills were overly reliant on less valuable but more plentiful hemlock. These hemlocks were harvested a decade later from the tree farm.
Mission Community Archives 0305-5

These higher-output mills sourced timber from forests further inland, including near Mission. In a brief jointly submitted to the second Royal Commission on Forest Resources chaired by Chief Justice Sloan in 1955–56, several of Mission's mill owners penned their frustration that local logs were not going to local mills. Instead, log harvesters

> invariably have working agreements with coast sawmills leaving little scope for small sawmills on the Fraser River to purchase boom logs. Actually, in 1954, it was estimated that 200,000,000 feet of logs went past Mission under the tow on the Fraser River to coast sawmills.

The authors of the brief recognized that "in recent years with many of the coast sawmills enlarging their plants and striving for increased production, it is becoming increasingly difficult for local mills to secure logs."

To survive, they instead were logging "small areas and patch timber, or in other words doing a cleanup job so far as the forests are concerned."

Mission's mills were left to buy logs on the open market, including a lot of hemlock, a wood considered less desirable because it took so long to dry out. Demand was much higher for Douglas fir, but hemlock was far more plentiful, quick growing, and previously overlooked by crews harvesting higher-value species. Sawmills in particular struggled. McMahon's Mission City mill shut down indefinitely in 1965 as a result of shifting markets and insufficient supply. McMahon referred to Mission's forest sector as a "hemlock economy."

By 1958, Mission's remaining sawmills were in a free fall. The three largest sawmills were McMahon's and Albert Thomas's mills in Hatzic, and Braich's on the Mission Flats. A decade later, two of these had shut down due to fires. McMahon held on to his mill in Dewdney until fire tore through the facility in 1984.

Those mills that survived did so by concentrating on producing cedar shakes and shingles. Toke Meeker and Henry Friesen established Anglo American Cedar Products Ltd. in the mid-1950s. The company remained one of the few from that era operating in Mission and Ruskin well into the new century. (It has since closed.) Watkins switched over to shakes and shingles in the 1970s. The Waldun Group bought out Stave Lake Cedar and operates another mill near the Stave River. The Cedar Shake and Shingle Bureau, a non-profit trade group, long maintained its main offices on the Mission Flats.

The Mission Tree Farm came too late to shore up lumber manufacturing in the District. Even after the tenure was established, Mission's relatively small mills required more timber than the municipal operation could provide in its first decade. The Mission Tree Farm would, however, spearhead the replanting and improvements to the land base needed to provide revenues and harvesting opportunities over the long run.

The Mission School Forest Association

FROM THE START, the District had envisioned that its municipal forest operations would support education, training, and research. The establishment of the Mission Forest Reserve in 1948 provided the first opportunities in inventory and restoration efforts. The reserve could be used as an experimental space for testing replanting and management strategies. Most of the reserve lands were located along or close to Dewdney Trunk Road in Steelhead, in areas easily accessed by the public, including those arriving on long yellow school buses.

Beginning in 1953, a group of civic leaders and educators met to plan for a school-run forest, an "outdoor classroom" that would help train the next generation of Mission's loggers and foresters. The Mission School Forest Association was established. Founding members of the MSFA included Albert McMahon of McMahon Lumber Company, who chaired and later served as president, and Albert Thomas of Thomas Lumber. Lang Sands, publisher of the *Fraser Valley Record*, was there. So, too, were educators from Mission Junior-Senior High School and representatives from the Faculty of Forestry at the University of British Columbia (UBC), the Canadian Forestry Association's BC branch, and the BC Forest Service. William Matheson, reeve for the District of Mission, and A.C. "Alf" Buckland, Mission's first municipal forester, also attended MSFA meetings.

The school forest hosted students learning about forest succession and planting techniques, as well as professional foresters, conservationists, and others interested in Mission's experiment.

The committee initially applied to the province requesting permission to operate a school forest on Crown land. It later withdrew its application after learning that the District was applying for a tree farm licence. In the interim, the MSFA identified, mapped, and moved onto thirty-five acres (later reduced to twenty acres) of municipally owned land on Bear Mountain, near the community of Steelhead. The parcel was close to town and part of the District's forest reserve. Students would learn forestry skills and plant management on land that had a diversity of soil types and even a small lake. The MSFA erected gates and signs to deter vandals, garbage dumpers, and the other unauthorized users that might be expected so close to town.

The school forest plot would support a newly developed forestry course, also a first in the province. In the classroom and in the field, the students studied current sustained yield practices and the reproduction, growth, and development of trees native to the Mission area. The school forest, as it was called, hosted students learning about forest succession and planting techniques, as well as professional foresters, conservationists, and others interested in Mission's experiment in community forest management.

This experiment in school forestry lasted for at least a decade, and likely longer, but official records on the school forest have been lost, discarded, or otherwise undiscovered. The MSFA dissolved at some point, but when is unknown. Perhaps the school curriculum changed and student interest in the forestry course dwindled, or the school forest became too expensive to maintain. By 1970, the municipal tree farm had absorbed the school forest lands.

Nearly sixty years after Mission established a training forest on Bear Mountain, UFV students familiarize themselves with the tree farm. *Alexandrah Pahl · all rights reserved.*

Mission's choice to pursue a TFL was a risky one.

The licence was designed primarily for large logging companies that could afford to take on expansive tenures.

Application

MISSION REALIZED THAT for a community forest to be successful, the municipality would have to find a way to take on management of nearby Crown lands. Without these lands, the forest reserve could not sustain itself, financially or otherwise, in the long run. In 1955, Mission applied for a tree farm licence so that it could expand into the more than 17,500 acres of Crown forest within municipal boundaries.

Mission's choice to pursue a TFL was a risky one. The licence was designed primarily for large logging companies that could afford to take on expansive tenures and work to maximize their yields and reforestation. The province instead encouraged local harvesters and communities to join together under a public working circle. Under this type of tenure, the BC Forest Service would be responsible for managing the forest base, and the timber would be sold off through a competitive bidding process. However, the creation of public working circles in the Fraser Valley often failed to materialize, even when mill owners and logging outfits pressed for them.

Alf Buckland, Mission's municipal forester, recommended in 1953 that a new management plan be developed for the timber in the District's boundaries. This represented the first step in formally applying for a tree farm licence. The District and the province reached an agreement-in-principle in 1954 that allowed Mission to move forward with a TFL application.

The province gave preliminary approval to the District's submission in early 1955. Mission's TFL application then ended up in bureaucratic purgatory, following the convening of yet another royal commission into BC's forest management. Chief Justice Sloan again led the commission, undertaking a review of the changes brought in following the previous inquiry. To the frustration of the District and some provincial officers, the Ministry of Forests put Mission's application for a TFL on hold indefinitely, pending the results of the commission's work.

The District of Mission tendered a brief to the second Sloan Commission in January 1956, in which it outlined the community's commitment to long-range forest health. Sloan's report, issued in 1957, included a detailed and generally positive ten-page overview of Mission's efforts to date. Mission's experience compelled the commission to make further recommendations to support community operations, including allowing access to Crown land outside of municipal boundaries. It also recommended that other communities interested in community forestry consider using Mission's management plans as a template.

The BC Legislature adopted many of Sloan's recommendations in early 1958, freeing up the ministry to make a final decision on Mission's application. By summer, Mission had been issued TFL 26, the first licence granted by the province to a municipality.

The nearly five-fold increase in size of the forest base gave Mission greater access to higher-quality timber and a larger area in which to plan operations. Mission had committed to long-term forest improvement efforts, even going so far as to suggest in its brief to the royal commission that "for at least twenty-five years all money earned from the forest should be spent on forest improvement work." It had promised to manage the tenure for multiple uses and to build roads as needed. And with half the forest too young to commercially harvest in 1958, the District would have to find strategies to balance the demands of markets with the need to reinvest revenues back into forest health.

Licence

THE POST OFFICE commands the most prominent corner, at 1st and Welton, of Mission's downtown strip. Erected in 1935, the building features an art deco granite facade above the entrance and landing. It was here, on July 24, 1958, that Mission welcomed Queen Elizabeth's sister, Margaret.

Princess Margaret's arrival drew an expectedly large crowd of dignitaries and local residents that warm summer day. Under the late morning sun, the twenty-eight-year-old royal alit from her motorcar onto a red carpet laid out for the occasion. British and Canadian flags adorned the landing where the princess stood as she extended her right hand to exchange courtesies. Her white handbag hung from her gloved left forearm. She waved to and charmed onlookers. Then, with a schedule to keep, Margaret turned to walk the route in reverse—across the landing, down the steps, along the red carpet, and into her car again. Her entourage promptly drove off.

News of the event featured prominently in the July 30 edition of the *Fraser Valley Record*, so much so that it dwarfed all other municipal business that week. At the bottom of the same page, a short announcement read:

> Culmination of a long series of conversations with Victoria, documents have been received by Mission Municipality granting the municipality Tree Farm License No. 26 which greatly expands the original Mission Forest Reserve.

On steeper slopes, cables and towers were used to pull cut logs uphill. *Mission Community Archives 0305-6*

That was it. After a decade of lobbying provincial officials, the news that TFL 26 had been awarded passed with little fanfare and no red carpet of its own. No perfunctory ribbon cutting or ceremonial signing marked the occasion.

Despite the limited reporting on the issuance of TFL 26, Mission's efforts to create a community-run operation had been well documented in the *Record*. The paper's editor and owner, Lang Sands, promoted healthy forests and the jobs that relied on them. In 1944, for instance, Sands published his detailed account of travelling to Vancouver Island with a delegation of political leaders from the Fraser Valley to tour forests replanted by the Alberni Pacific Lumber Company.

A decade later, the *Record* began printing an annual special section on Forest Conservation Week. Sands lobbied local forest products

After a decade of lobbying provincial officials, the news that TFL 26 had been awarded passed with little fanfare and no red carpet of its own.

manufacturers for donations and ad sales which would pay for the section and the "Keep BC Forests Green" campaign. For years, it was only one of two editions of the paper when colour—green, specifically—was added to part of the paper. The other colour edition ran during the Christmas season.

Through the *Record*, Sands built support for locally managed forests. The paper ran the necessary forestry-related stories on summer droughts, mill openings and closures, and labour strikes. But Sands also reproduced, often in full and on the front page, the reports and management plans and speeches made to the board of trade and Mission council in favour of community forests and public working circles.

The awarding of TFL 26 marked the culmination of efforts from within Mission and the forestry industry over a period of fifteen years. During that time, Mission designated its own forest reserve, lobbied Victoria for changes to the Municipal and Forest Acts, hired a municipal forester, developed a forest management plan, and even created a dedicated school forestry operation. Mission was undertaking the most ambitious community-led forest project yet to be seen in BC.

6

A GROWING OPERATION

BEFORE TAKING THE reins as Mission's first tree farm manager, Irwin "Rocky" Rockwell served as the provincial ranger for the Mission area. In his new role, he managed the day-to-day operations and oversaw longer-term planning. It was also Rockwell's job to provide reassurance to the District and province that the tree farm experiment might just generate a return on the municipality's investment.

"The real value, of wages earned by contractors and other workers, is difficult to assess," Rockwell wrote in the 1959 annual report, "but there is no doubt that through the years, it will be a very tangible benefit to the stability of the economy of the District."

PRECEDING Touring the tree farm, late 1960s. *Mission Community Archives 0305-7*

FACING Cones protect recently planted one-year-old western red cedar seedlings. *Kelly Cameron*

Over time, this "benefit" took on many forms, including jobs for locals and expanding public access. In that first decade, however, Rockwell and his crews concentrated on the essentials. District staff and contractors cut overgrown second-growth forest and salvaged cedar, planted millions of seedlings, bulldozed miles of new roads, and extracted valuable Douglas firs to pay for it all. In doing so, they transformed the cutover lands of north Mission into a viable, sustainable working forest with the potential to generate revenues for the District for years to come.

With the tree farm licence, Mission took on the cost of dozens of miles of new road in the first decade alone. *Mission Community Archives 0305-8*

The First Managers

ANY NUMBER OF things could have undermined the political and ecological viability of Mission's experiment. Fires, infestations, or prolonged drought. Market crashes, disputes over vision and decisions, or budget shortfalls. The previous years spent operating a forest reserve had provided some sense of the challenges that lay ahead. But the forest reserve was much smaller than the TFL, and most of it was fragmented and adjacent to private homesteads.

Instead, those things that went right helped to ensure the tree farm's success. Arguably, no other factor was as important as the tight-knit network of local forestry professionals hired to work on the forest reserve and TFL. In 1953, engineer A.C. Buckland was brought on as Mission's first municipal forester, after working in the private sector. He assessed Mission's reserve lands for timber values and undertook needed planning. Buckland eventually left Mission to take a job with Canadian Forest Products in June 1961.

Buckland was followed by BCFS ranger Jim Robinson. Robinson had overseen the Mission District of the BC Forest Service, which included the Crown forests north of the Fraser and east of the Pitt River. From his office in a two-storey white clapboard ranger station on the west side of town, Robinson managed a crew of five men in the early and mid 1950s. Working under him was Rocky Rockwell, who took a job as an assistant BCFS ranger in 1950.

Robinson and Rockwell were co-workers, friends, and neighbours. The two met after the war, while waiting in line for a veteran's land grant. According to Rockwell's son, Lorne, his father was waiting to sign the paperwork to purchase ten acres in north Mission with his grant allocation. Robinson was standing behind Rockwell and struck up a conversation. The two agreed on the spot to split the ten-acre parcel and its three-hundred-dollar price tag.

While with the BC Forest Service, Robinson and Rockwell assisted Buckland on the Mission Forest Reserve. Robinson soon moved on to the Vancouver Forest District in 1954, leaving Rockwell to take over as the BCFS ranger-in-charge for the Mission District. Robinson returned to work in Mission briefly, when in November 1958, the Mission council named Robinson to be Buckland's replacement as municipal forest ranger.

Shortly afterwards, Rocky Rockwell replaced Robinson as head of the Mission Tree Farm. Rockwell held this post for two decades, leaving an indelible mark on the tree farm. The tree farm gave Rockwell the space to practise innovative forestry, including controlled burning trials and silvicultural experiments in yellow cedar reproduction and cultivation. Just as important to Rocky's wife, Mary, the job kept him close to home in Mission, where they raised their family.

At the start, Rockwell managed the day-to-day operations with the assistance of municipal clerk E.E. Chace and his secretary. Like many rangers of his generation, Rockwell had trained at a government facility in Surrey called Green Timbers, but he did not have the courses needed to become a registered professional forester (RPF). Rockwell worked closely with Ian Scheidel, an RPF who worked on contract for the District. Scheidel signed off on paperwork for which the province required RPF review, worked out the placement of new roads built across the tree farm, and surveyed areas for harvesting, or cutblocks.

Rockwell and his staff reported to the Forestry Committee of Mission's municipal council. The committee usually consisted of the reeve and two council members. Invariably, the membership had close ties to the local timber industry. In the 1950s and 1960s, the public largely deferred to

Small crew installing a plank road, late 1960s. *Mission Community Archives 0305-9*

these elected representatives to be their voice in the daily running and longer-term planning of the tree farm.

Mission, still a small community, was starting to face the challenges of growth. Local interest shifted to other issues of pressing importance—a new hospital building, for instance, and the need for a dedicated vehicle bridge over the Fraser River. (Prior to 1972, vehicles used the train bridge to cross.) The tree farm's management, meanwhile, would be folded into the local government structure, operating as Mission's Forestry Department.

Mission Tree Farm manager Irvin (Rocky) Rockwell leads UBC forestry students on a tour, late 1960s. Mission was BC's sole working community forest operation at the time. *Mission Community Archives 0305-10*

The First Years

THE FIRST FULL year of operation, 1959, saw a harvest of more than 6,411 cubic metres of hemlock, fir, cedar, balsam, white pine, and yellow cedar for use as saw logs, plus another 959 metres of cedar poles and 56 cords of cedar bolts. Municipal staff and contractors also removed salvage materials to be used as shakes, bean stakes, fence posts, and pulp.

By comparison, in 2013, Mission's forestry crews and its largest contractor, H&C Logging, harvested more than eight times that volume—nearly 45,000 cubic metres of saw logs, poles, pulp, and cedar bolts.

Manager Rocky Rockwell wrote of the growing pains of getting the tree farm up and running that first year:

> Problems of access, need of heavy capital expenditures on equipment, underproduction due partly to the logging strike, insufficient knowledge of the resources of the Tree Farm License, and other difficulties, all combined to point out the fact that we have much to learn and more to do in order to get started in the right direction.

Rockwell and forest engineer A.C. Buckland focused on essential planning needs in 1959–60. Their top concern was the risk of fire; they developed fire prevention and response procedures on TFL 26 and stocked up on fire equipment needed to suppress any flare-ups. Crews erected a weather station in Steelhead and collected readings every day, twice a day, which they reported to the Ministry of Forests.

Mission's persistence in establishing a community forest made it a visible and national symbol of progressive change.

Meanwhile, Rockwell's team directed some of their efforts towards improving accessibility. Municipal crews pulled out a rotting plank road west of Stave Lake and used tractors to put a truck road in its place. In those early years, however, Mission's barebones forestry team lacked basic equipment and office furniture. The forestry team did not even have vehicles that Rockwell could use to access the tree farm. He had to borrow a truck from the District's public works department.

All of this served as a reminder that the town was building a community-run operation from scratch. There was no template to draw upon. The District had taken on managing second-growth Crown forest, not virgin timber. It had done so at a time of enormous change in the industry and under a type of licence that had not been designed with small towns like Mission in mind.

Rockwell, Buckland, Robinson, and Mission council's Forestry Committee were also acutely aware that what they were doing was being watched by many outside the District. Local MLA Lyle Wicks, who also served as the minister of labour and railways, was one the tree farm's vocal supporters. Speaking before the BC Legislature in 1959, Wicks held up the Mission Tree Farm as an example to be followed by other municipalities. West Vancouver considered pursuing its own tree farm licence, based on the success of Mission's venture.

A truck from a local firm transports a load of cedar bolts down the West Stave Access Road. *Mission Community Archives 0305-11*

Mission's persistence in establishing a community forest made it a visible and national symbol of progressive change. The Canadian Forestry Association lauded the District's efforts in conservation early on, when it named it the first-ever Forest Conservation Week Town in 1951, and when it organized a conservation outing to the reserve in 1956. In August 1960, CFA president Ben Avery recited "The Mission Story" before foresters from around the world attending the Fifth World Forestry Congress in Seattle.

All this attention would be for naught if Rockwell and his team could not address the challenges that lay ahead of them. They had to find ways to increase the annual cuts, expand the road network, deal with the demands created by wood waste, and ensure successful replanting. Perhaps most importantly, they had to generate revenue from harvesting to cover the costs of the whole operation.

Harvesting

TREE FARM LICENCE in hand, the District put the first parcels of timber and salvageable cedar up for sale in late 1958. Interest from local contractors was high. Three of the four cutting permits awarded went to Herman Braich's sawmill, with wood to be processed along the Mission waterfront.

Timber sales were, then as now, determined in line with the District's management plan for TFL 26. The provincial government requires licensees to submit a plan as a condition of maintaining and renewing their tree farm licences. Mission's first management plan applied to more than eighteen thousand acres of forest. Approximately three thousand of these acres were owned outright by the District (Schedule A lands); the rest were Crown (Schedule B) lands.

Schedule A and B lands were managed together according to the principle of sustained yield management. This was achieved in part through setting an annual allowable cut limit. In 1958, the province set the AAC for TFL 26 at 12,035 cubic metres per year for each of the first five years of operation. Mission could vary its annual cut, so long as the five-year average was within 10 percent of the AAC five-year target.

Decisions on where to harvest are based on age, species, and accessibility. Age of the trees is paramount, as older trees usually provide more commercially valuable wood. In 1958, most of Mission's forest could support some commercial timber extraction, although half of it was under fifty years of age and still decades away from being harvestable. Just over a fifth of the forest was old growth.

In the early years of the tree farm, harvesters logged and salvaged Douglas fir, yellow and western red cedar, hemlock, pine, and small volumes of other species. Fir commanded the highest price in the market, and so Mission replanted fir seedlings almost exclusively. Much of the best timber was off limits because of insufficient road access. Mission would take more than a decade to reach the timber in the northwestern quarter of the tree farm.

More or less the same procedures have been used to sell wood from the tree farm since its establishment. Once marketable stands of timber were selected, a forest engineer would lay out (or outline) the cutblocks to be harvested. For most of the 1960s, this work was contracted out to a local forester, Ian Scheidel, RPF. Next, a timber cruiser would provide an assessment of the volume, quality, age, and species in an area to be harvested.

On all other licences, the measurements collected would be used to determine stumpage rates, paid to the province for the right to commercially extract timber. For its first two decades, however, Mission's tree farm benefited greatly from an arrangement in which it paid royalties in lieu of stumpage.

The District's forest crews did only a fraction of the commercial cutting themselves. As a requirement of the licence, at least half of the harvest from Schedule B lands was to be undertaken by contractors rather than municipal staff. Once a year, cruised blocks of timber were put out for tender. Usually, three or four blocks of varying size were put up for auction each year. The District also took bids on cedar salvage. Competing bids were then reviewed by council. (Today, they are reviewed by the Forestry Department.)

Once the trees were cut, they were shipped by truck to a log-sort yard close to the Fraser River. There, a timber scaler (trained in doing evaluations and calculations) estimated the quantity and value of wood. Saw logs were measured and recorded so that sawmills could determine the volume of lumber to be produced. Trees logged for poles were measured in linear feet or metres. Cedar used for shakes and shingles was measured in cords.

From layout to salvage to planting, the District historically directed harvesting work to locals as much as possible. Until the past few years, Mission prioritized local tenders and contracts over those from outside the District. Early timber sales were deliberately kept modest to encourage bids from independent operators. Local contractors were also hired by the District for jobs too small to put out to tender. Well into the 1970s, municipal crews, including seasonal hires in the Winter Works employment program, completed much of the cedar salvage work and almost all replanting.

District of Mission forestry workers did many jobs too small or unprofitable for contractors, including clean-up, salvage, and some harvesting. *Mission Community Archives 0305-12*

Decisions on where to harvest are based on age, species, and accessibility. Older trees usually provide more commercially valuable wood.

Many of the same processes for cruising, harvesting, and scaling are used today at increasing speed and scale. Computers improve processing times, accuracy, mapping, and valuation. Forest managers benefit from high-resolution satellite imagery and LIDAR, a system that uses lasers to collect landform data for building 3-D maps. Timber cruising and other tasks are contracted out to credentialed specialists who use digital tablets and equipment to lay out cutblocks.

Critically, more care is given today to completing environmental and archaeological inventories and values before a single shovel hits the ground. There is no telling how much of the traditional Stó:lō history of the landscape was lost to early forest activity, before and after Mission undertook TFL 26. The building of road networks alone, including west of Stave Lake, cut through important settlement and resource sites.

The seasonal workforce that once completed much of the manual labour in the forest is also gone, as Mission residents are less likely today to work in the woods. Tellingly, local companies are no longer given preference in timber sales, in part because there are fewer local operators. Following a drawn-out crisis in the forest industry during the 1980s, and with the availability of more agile and powerful harvesting equipment, more logs are cut today by fewer and fewer people.

The Fire Fighters

UP UNTIL THE 1970s, the prevailing logic in forest management called for putting out fires as quickly as possible. Fire suppression was viewed as necessary to protect communities and valuable timber. Ecologists and foresters questioned the value of fire suppression, but it would take much longer for communities and government agencies to accept the risks that came with letting more fires burn.

When they did, it was because smaller, more frequent fires shape healthy forests and reduce risks of bigger fires over time. A low-intensity fire will creep along the forest floor, feeding on the woody remains of fallen trees and broken branches. Like waves against a breakwater, flames will lap against the thick bark of mature conifers. Most healthy, older Douglas firs, pine, and spruce trees will come through the fire okay. So, too, will many of maple, balsam poplar, and other deciduous trees.

Frequent fires prevent the build-up of fuel, the word fire engineers use to refer to dead and dying organic materials in a forest. Fire reduces competition, burning younger trees that compete for water, nutrients, and sunlight. Fire changes the soil pH and makes nutrients more accessible to seedlings. Many species thrive after fire, and some depend on fire for propagation and survival.

This is true even in BC's wet coastal forests. The Stó:lō and their ancestors that lived within the Stave watershed once used fire as a management

FACING Prescribed burning near Hoover Lake, 1968, part of Canadian Forest Service's study on fire management. *Mission Community Archives 0305-13*

tool. They lit small burns to encourage berry production and the growth of grasses that attracted game. Periodic fires made it easier to travel through the area's dense forests.

Fire swept through the forests of the Stave Valley several times during the eighteenth and nineteenth centuries. Up to 90 percent of the tenure burned during fires in 1845, 1880, and 1927. By the 1950s, though, forest fires created a much greater danger than they had prior to the arrival of settlers.

More people lived at the forest's edge, which added to the risk. The bigger culprit was a half-century of logging that was both under-regulated and grossly inefficient. Early logging crews left almost as much behind as they took out. Wood processors had no interest in "slash"—the broken trees, treetops, sawdust, lopped-off branches, and wood chips left over from logging, except as fuel to heat some of Mission's homes in the winter. In the 1910s through to the early 1930s, logging outfits harvested the fire-resilient mature timber of Mission's forest but did not plant replacement stock. The naturally regenerated second growth was patchy, overgrown, and entirely unmanaged.

Across the Vancouver and Mission Forest Districts, the timber industry created a highly combustible topography. Slash piles and scattered wood waste became waterlogged from the winter snows and spring rains but dried out during the summer months. The immature second-growth forest could not shade out new and prolific plant growth. A hot enough fire could easily have torn through a dense stand of twenty-year-old hemlocks and cedars.

Other areas of the province had already experienced slash fires. In 1938, almost seventy-five thousand acres burned between Campbell River and Comox, on Vancouver Island, after a fire broke out in logging waste. Fires in other logged-over areas happened frequently enough that the provincial government changed the Forest Act in the 1930s to require crews to burn their slash piles after harvesting. Fire science, though, was still in its infancy. *How*, *when*, and *where* to burn was poorly understood, and crews were often unwilling to start new fires.

Fire reduces competition, changes the soil pH, or acidity, and makes nutrients more accessible to seedlings. **Many species thrive after fire.**

Support for better forest management came from all corners, including from the logging industry itself. At the 1952 annual meeting of the BC Truck Loggers Association, president J.W. Baikie told the crowd, "The public should be informed that fire is the number one threat to growing trees as a crop." In the 1920s, the Canadian Forestry Association founded North America's first Junior Forest Warden program in BC, with the express purpose of training in fire prevention and forest protection. By the mid-1950s, thousands of boys and young men across the province, including many in Mission, had taken part.

• • •

In the Mission Forest District, the BCFS ranger station cautioned the public and industry to remain particularly vigilant during the summer months. The Forest Service office decided when to close the forest off to public access, which would occur most years. The choice was never an easy one, because closures caused loggers to lose work and families to go without a paycheque.

It all depended on the weather. Some years, the forest was closed as early as the first or second week of July. In 1944, the Forest Service closed the forest on July 13 and yet still recorded 211 fires in the Mission District of the BCFS a week later. The following year, the forest stayed open through the summer, but that September, fire broke out in old slash on the west side of Stave Lake. The fire destroyed equipment, a steam donkey, and more than two hundred thousand logs owned by a company that was operating there.

During the summer of 1951, loggers were out of work for almost two months due to the fire risk, until rains returned at the end of August. A week later, the closure was back on, the forest having been open only long enough for campers to take advantage of the Labour Day weekend.

In July 1958, when conditions were "dryer than dry," according to one official, lightning strikes sparked several forest fires in Mission. Municipal and BC Forest Service crews were monitoring weather conditions and working to put out fires near Alouette and Stave Lakes. The official noted, "Even the humus in the ground is burning now."

Only a week earlier, the province had awarded Mission the tree farm licence. Mission quite literally was going through a trial by fire.

• • •

Most of the timber on the Mission Tree Farm in 1958 was classed as immature, or less than seventy years old. The operation's first management plan, drafted two years earlier, identified more than 3,600 hectares that was less than thirty years old. Much of the forest was young and growing on top of decades-old rotting slash.

Fire risk mitigation took precedence, and harvesting, thinning, planting, and road building were done with an eye towards reducing the hazard. The BC Forest Service directed fire suppression work, while the District organized a fire response protocol. Some of the early years' revenue was spent on fire equipment, including a truck-mounted water pump and hoses. The District agreed to pay for expanded fire patrols in the summer and for weather monitoring and reporting.

Mission's forestry crews reduced fire risks by aggressively rehabilitating second-growth forest, including that classified as non-commercial because of its poor condition. The District worked with contractors, municipal departments, and the BCFS to improve fire response readiness. In the spring and fall, municipal crews and seasonal workers set controlled fires to burn off the stumps and branches left behind in recently logged areas.

Fires continued to break out *near* the tree farm, including one at the head of Stave Lake in 1960 that burned 240 acres and more than 750,000 board feet of cut timber. But the tree farm enjoyed a fortuitous first decade free from accidental fires.

Rocky Rockwell expressed measured satisfaction with the tree farm's successes in fire prevention. "We know that they can't go on forever," he told a reporter from the *Vancouver Sun*, "but we can sure try to keep it going on as long as possible."

Investments

IN 1960, IN only its second year of operation, the Mission Tree Farm made a profit. Sort of.

That year, the tree farm doubled the volume of wood harvested over the previous year's total, to 13,260 cubic metres. The District reached its AAC and expanded road access. A green crew of planters averaged just over three hundred juvenile trees per person, per day, planting more than 35,500 Douglas fir, yellow pine, and poplars. The District ended the year a few thousand dollars in the black.

The District had to spend money to make money, and surplus revenues were used to improve the viability of the tree farm. In 1958, neither Mission nor the province had a detailed count of the type, age, and value of trees on the tree farm. Parts of the forest were deemed non-commercial, but that was an assessment based on current conditions at the time and not on future possibilities. On top of this, much of the tenure lacked road access. Revenues were used to survey road right-of-ways and conduct inventories of forest stands. Salaries ate up most of the budget, as crews were hired to cruise timber, pile and burn waste, spray and pull weeds, and replant harvested areas. Additional funds went to buying fire equipment and building fire trails and breaks.

As annual volumes of cutting went up, so did revenues. By the late 1960s, the tree farm was harvesting anywhere between fourteen thousand

and twenty-five thousand cubic metres of timber each year. In 1969, almost 26,600 cubic metres came off the tree farm. The District netted one out of every three dollars coming in from timber and salvage sales.

Early investments shored up needed political support within the community. In the years since Mission's business leaders first embraced the notion of a community forest, several of the early champions had retired or passed away. And the District council had already spent nearly ten thousand dollars on an experiment in community forestry that lacked precedent or guarantee of success. So far, the reserve lands had generated only limited returns, and none that could be used to offset local tax burdens.

One of those expressing skepticism was F.R. Hall, who later became reeve in December 1963. Hall admitted to having "never been particularly enthusiastic about the Mission Municipal Tree Farm" before taking office. His decision to run for a seat on council was in part because he felt that Mission was losing out on opportunities to develop the area.

Politically, the most important spending decisions involved hiring laid-off workers. In Mission, seasonal unemployment in fishing and forestry occurred frequently due to winter weather and drought-related closures during the summer. The first to receive pink slips were the logging crews, but as mills worked through their stockpiles, they, too, sent workers home.

Rocky Rockwell, the tree farm's manager, had plenty for them to do. He hired upwards of fifty men each year to clear brush and thin forests, clean up waste, recover salvageable wood after logging, and replant trees. Some of those hired had struggled to find stable employment for years.

Soon after he became reeve, Hall's attitude towards the tree farm changed. He expressed satisfaction with how much wood was coming off the tree farm and that surpluses were being generated. Hall was especially enthused about the number of jobs being created. "Last winter we were able to practically use all unemployed persons through either our Winter Works Program or through expansion of our Social Assistance Program," Hall noted in a 1964 interview to the *Fraser Valley Record*. "They were given employment on the Tree Farm and the products of their labor offset their actual cost to the municipality."

Mission's costly investments in the tree farm during its first decades included new forest access roads west of Stave Lake and up to Hoover Lake. *Mission Community Archives 0305-14*

Reeve Hall was not the last elected official in Mission to question the value and wisdom of maintaining a municipal tree farm. Support has always been predicated on some type of return to the District, such as jobs, revenue for the general coffers, or hiking trails and boat launches. There have also been those whose opposition to a municipal tree farm is more ideological, grounded in a belief in small government or a preference for private enterprise.

Every tree farm manager encounters skepticism and confronts the challenge of time. The types of investments needed to sustain a forest operation do not typically pay off in a political cycle. Especially in land acquisition, planting, and stand management. When markets were healthy and prices were high, the tree farm funded the three Rs: roads, rehabilitation, and replanting. But the returns on these investments were measured over decades.

Throw into the mix the cyclical nature of timber markets, and it can be easy to see why political support can wane. The demand for BC wood products is tied closely to western Canadian and US housing markets, and the industry experiences periodic slowdowns. Postwar suburban growth created a surging demand for shingles and fencing, and the 1960s was a good decade for the province's timber industry. Then in 1970, prices dropped so low that, when Mission put three timber blocks out for tender, or auction, there were only a handful of bidders.

Rockwell was also forthright about Mission's biggest advantage in those first decades—stumpage, or rather the lack of it. The original Tree Farm Licence 26 was issued for a twenty-one-year period and would be renewed so long as licensees submitted regular management plans and met AAC targets. Private firms with TFLS paid a fee on each tree cut, indexed to market values. In a one-of-a-kind arrangement, TFL 26 instead required Mission to pay royalties once a year. The difference was substantial. Rockwell estimated that the resulting savings averaged fifty thousand dollars annually during the first two decades of operation. Rather than being used to cut a cheque to the province, this money was directly reinvested into the tree farm.

Timing was everything. Had the tree farm started a decade later or had Mission been on the hook for stumpage rather than royalties, the fate of the enterprise—politically and financially—might not have been the same. Rockwell anticipated that this arrangement would someday change. In the meantime, he and his team forged ahead with improving the odds that the Mission Tree Farm would remain fiscally and ecologically viable.

The rural community of Steelhead is home to much of the municipally owned forest land in the tree farm. *Justine Robinson · all rights reserved*

Lands

IN 1963, REEVE F.R. Hall and Mission's council met with the BC minister of agriculture, Frank Richter, to discuss a plan to acquire new lands under the provincial Agricultural Rehabilitation and Development Act (ARDA). Hall saw the tree farm—and ARDA funding—as a possible solution to a long-running problem.

Generations of settlers had toiled in what were largely "futile attempts at farming," according to the District. "Since the 1920s, the Steelhead area has a history of distress." Abandoned and underperforming Steelhead lands had become an economic burden to the municipality, with fewer residents and declining tax rolls. The ARDA plan called for the purchase of 1,600 acres tied up in forty-three properties over a five-year period, with the land to be added to TFL 26.

This was not the first time the District sought to acquire land from private holders. In early 1944, the District had attempted a land purchase for a reforestation project, but to no avail. Once the tree farm was up and running, the District had more success. Multiple times in the tree farm's first decade, Rockwell worked with other government departments to locate and purchase additional land for the TFL. Most of this activity was in the Steelhead area. Seventy acres were bought and added to the tree farm in 1961, along with thirty acres of municipally owned land. Forestry revenues paid for another 142 acres the following year.

Land acquisition helped plug the holes along the periphery of the tree farm. Privately owned lands abutted or even surrounded municipally owned lands in the tree farm.

Land acquisition helped plug the holes along the periphery of the tree farm. Privately owned lands abutted or even surrounded municipally owned lands in the tree farm. The pattern of mixed ownership made building roads difficult in spots. "Obtaining these right-a-ways was found to be a long and tedious process, especially when prices asked are out of all reason," Rockwell wrote in the 1960 annual report. "The lesson learned here is that the Municipality must continually be ready to obtain access wherever needed even though they may not be required to in the near future."

Ever more emboldened, Reeve Hall—the one-time skeptic of the tree farm operation—even sought out forest lands in Garibaldi Park. Located to the northwest of TFL 26, Garibaldi is widely considered a crown jewel in the BC Parks system, a glaciated landscape of sawtooth peaks towering over the still waters of ice-fed alpine lakes. A southern arm of the park stretched south to the western shore of Alouette Lake and the northwestern corner of Stave Lake.

In 1967, this section of Garibaldi was rededicated as Golden Ears Provincial Park. Hearing of the provincial government's plan to develop

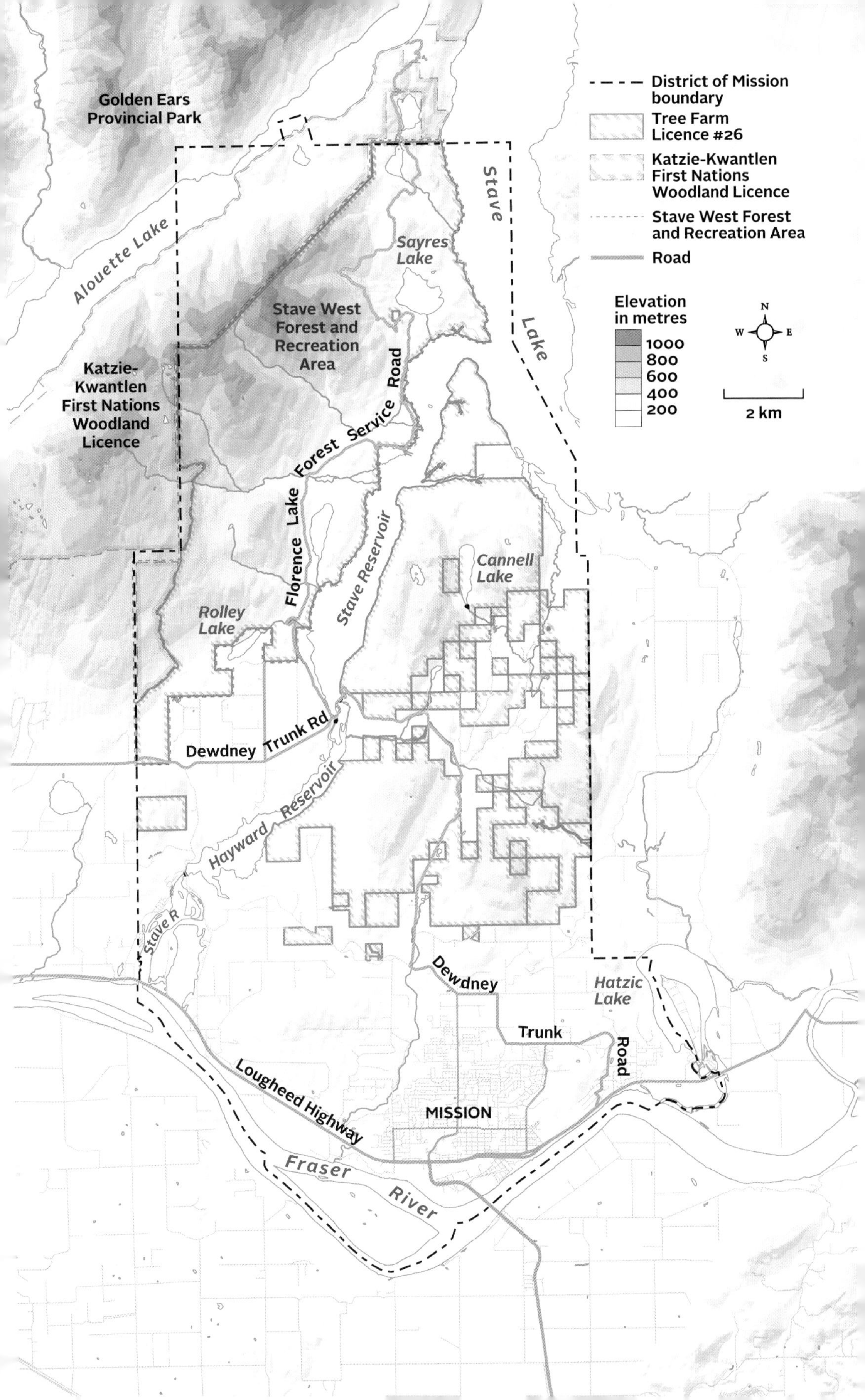

District of Mission boundary
Tree Farm Licence #26
Katzie-Kwantlen First Nations Woodland Licence
Stave West Forest and Recreation Area
Road
Elevation in metres
1000
800
600
400
200
N
W
E
S
2 km
Golden Ears Provincial Park
Alouette Lake
Stave Lake
Sayres Lake
Stave West Forest and Recreation Area
Katzie-Kwantlen First Nations Woodland Licence
Florence Lake Forest Service Road
Stave Reservoir
Cannell Lake
Rolley Lake
Dewdney Trunk Rd
Hayward Reservoir
Stave R
Dewdney Trunk Road
Hatzic Lake
Lougheed Highway
MISSION
Fraser River

facilities within Golden Ears and Garibaldi proper, Hall proposed extending the tree farm north into the park. There, it could access mature timber stands and also "make the area more suitable for recreation than it is in its present state." Hall's proposal died quickly.

Land was periodically added to the tree farm in the following decades. Most sizable additions resulted from requests for Crown land rather than through municipal purchase of private property. In the early 1970s, Mayor Neville Cox and his successor William Harris, along with the MLA for Dewdney, Peter Rolston, lobbied the provincial government to transfer nearly five thousand acres from old timber berths (or licence areas) into TFL 26. The timber berths in question were originally lands granted to the CPR in exchange for building the railroad. The CPR had sold the rights to timber to logging companies. Once cleared, the land reverted back to Crown control.

The original TFL lands surrounded and abutted many of these timber berths, which in turn meant fewer efficiencies in harvesting and road building. Plus, most of the berths were in poor condition. Rockwell acknowledged the challenge of folding the berths into the TFL. "The short-term value is, of course, not that big. Most of the areas have been logged but haven't been planted. They will have to be rehabilitated. But," Rockwell added, "in the long run it will considerably increase our cut, by about 20 percent."

The provincial government agreed. In 1974, it transferred 3,360 acres into TFL 26. The result was an increase in the TFL's total size by almost 18 percent. The municipality took control over lands near Hoover, Hayward, and Cannell Lakes. Land on the western edge of the tree farm was added, as was acreage west of Stave Falls Dam.

At the time, the province declined to transfer desirable lands near Stave and Cedar (later Sayres) Lakes, and these were kept in reserve for potential park development. Small but active timber berths held by older licensees existed within the boundaries of the TFL. And still more land adjacent to the tree farm remained just out of reach for the same reasons. The Mission Tree Farm had not yet finished growing.

Roads

WHETHER ONE IS moving by foot, train, or by truck, the usual goal is to find the path of least resistance to reach valuable resources. When commercial logging began in the Mission area, some of these routes were already well worn. Indigenous trails once connected the Stave River delta with berry and cedar collection sites and summer encampments on the lake. Plank roads followed some of these same Indigenous migration trails.

Mission's forest lands became more accessible when the province built the Dewdney Trunk Road in the late 1800s. The road connected Port Moody, east of Vancouver, with the rural community of Dewdney, east of Mission. Whonnock, Stave Falls, and Steelhead all grew up along the route. Today, the Dewdney Trunk Road skirts the southern periphery of the tree farm west of Stave Lake. East of the Stave Falls Dam, the road winds uphill in a series of hairpin curves, past the municipal dump and beyond Steelhead, threading between Bear and Red Mountains, and then into the suburban Cedar Valley neighbourhood of north Mission.

North of Dewdney Trunk, early logging outfits relied on skid roads built deeper into the forest. Remnants of some of these roads are still visible near Hoover and Devils Lakes. At Hoover, loggers once felled massive trees into the water and then floated the logs to a skid road to be hauled downslope. A century later, the road's planks lie split and broken, still visible but partially submerged in mud and water.

A road-building crew tries to plan the route along a steep slope, mid-1960s.
Mission Community Archives 0305-15

In the late 1950s, municipal crews reached stands of timber closest to Dewdney Trunk and by way of side roads in Stave Falls, Silverdale, and Steelhead. Dirt roads extended up the lower flanks of Bear and Red Mountains. Mission's priority was constructing well-graded access routes from which secondary logging roads could be built up to higher-elevation stands.

Trucks and logging equipment required hard-packed roads. The expansion of dirt and gravel roads was crucial to lowering the costs of harvesting and replanting. New roads shortened the time needed to respond to lightning-caused forest fires, even if more public access sharply increased the risk of human-caused conflagrations. Between 1959 and 1961, modest timber revenues paid for several miles of new roads, including one to the east of Cannell Lake to a BC Forest Service fire lookout.

• • •

Holders of tree farm licences were solely responsible for their own road construction and maintenance costs. That was just fine by Mission, as the municipality retained control over where and how the roads were put in. "Culverts go where they should and proper ditches are dug," explained Howard Murdoch, tree farm foreman (1968–80).

New road construction was one of the tree farm's costliest investments. Road building starts with logging the right-of-way. This job is done by the skidder operator, otherwise known as the cat skinner (a reference to Caterpillar tractors). A dozer follows behind to clear out brush, stumps, large rocks, and soils. Crews left stumps behind to provide structural support to new roads. They also added ditches and culverts to control and channel water downhill.

Dozers levelled out roadways and compacted surfaces. In the uneven terrain of the tree farm, crews built up roads against hillsides, often cutting into the upslopes. The dozers would move rock to the opposite embankment, where it would be repeatedly compacted to prevent slope failure. The process was time-consuming and often dangerous, even more so when crews had to cut or drill into hard rock.

Along most logging roads, bulldozers were sufficient to do the job, but later, excavators were brought in for larger, steeper, or more highly trafficked roads. Excavators have several advantages, including an ability to scoop materials into the back of waiting trucks. If the goal is to create a more stable surface by digging down to bedrock, they are more effective machines. The tree farm, however, did not purchase its first excavator until the early 1980s.

The road network spidered across lower elevations of the tree farm. As prices for yellow cedar and fir soared in the late 1960s, the District could afford machines of increasing size and power, and roads could be put in more quickly. The process for building the roads, though, remained much the same. In the wet climate of the north shore, the stumps entombed in roadbed degraded more quickly, allowing water to seep in. Earlier bulldozed roads were prone to washouts and slides.

• • •

In September 1968, the BC Forest Service reported that another forty kilometres of road would be needed on the tree farm over the next quarter century. In order to meet this target, the District needed to add three kilometres of new road per year between 1969 and 1974—double the rate of the previous decade. Rockwell had estimated at the time that it would take another twenty years to complete the tree farm's road network.

Nonetheless, Rockwell assured council that they were on the right track. "For the first time since the tree farm license was granted," he wrote, "there is now enough new road in to give access to more timber than is needed for next year's cut, especially in the higher altitude old growth timber."

Road building accelerated, consuming most of the tree farm revenues for several years to come. Improvements were made to the Hoover Lake Road. New branch roads provided access for fire prevention and response and opened up new stands of timber. In 1971, Rockwell had budgeted for just under five kilometres of new road, the most in one year yet. The pace slowed to three kilometres each year by 1974.

By this point, the completion of the sixteen-kilometre West Stave Access Road (WSAR) on the western half of the tree farm had created access to large stands of high-value timber. Of all the roads built in the 1960s and 1970s by the District, the WSAR was by far the most complex, challenging, and at times frustrating project, but it also paid out tremendous dividends. The WSAR and the secondary roads that were subsequently built off it enabled Rockwell and the District to plan forest operations and generate revenue on a much larger scale.

Dozers levelled out roadways and compacted surfaces. **Crews built up roads against hillsides, often cutting into the upslopes.**

The West Stave Access Road was Mission Forestry's most expensive infrastructure project. Crews finished this section just south of Rolley Creek in 1963. *Mission Community Archives 0305-16*

West Stave Access Road

MISSION ASSUMED RESPONSIBILITY in 1958 for forest lands that no one from the District's council, no member of District staff, had ever visited. The forests west of Stave Lake and miles to the north of the Dewdney Trunk fell within municipal boundaries but were otherwise quite remote. Footpaths once criss-crossed the forests between Alouette and Stave Lakes, used by Indigenous families for summer berry collection and hunting. But the difficult topography deterred most settlers, and the northwestern corner of the tree farm had been visited of late only by the occasional cedar salvager, surveyor, or treasure hunter.

Plank roads were once constructed west of Stave Lake as far north as Devils Lake, but north of this point, the wilderness that flanked the eastern side of Mt. Crickmer was unroaded. And with good reason. Opening the west side of the tree farm to logging entailed constructing a route across challenging terrain and several large water courses. Without a modern road, however, forestry crews would not be able to harvest the large stands of valuable old-growth cedar and fir found north of Seventynine Creek. Thus began the single largest infrastructure project in the history of the tree farm—the building of the West Stave Access Road.

The WSAR came to be known locally as the Burma Road and later (officially) as the Florence Lake Forest Service Road (FLFSR). Site preparation for the WSAR began in 1961. The road paralleled the western shoreline of Stave Lake as far north as Kearsley Creek before veering inland to the west of Cedar (now Sayres) Lake. During the first year, municipal crews cleared a half-mile (0.8 kilometre) right-of-way. They inched closer to Rolley Creek the following year and erected a wooden bridge by November 1963.

Crews reached the four-mile (6.5 kilometre) point of the WSAR in 1964, and surveyors mapped ahead two more miles, up to Seventynine Creek. Then progress slowed considerably as crews traversed more difficult terrain. In November 1965, as a Caterpillar D8 tractor slowly cut its way across an old slide area, the hillside gave way. The tractor ended up four hundred feet below. Only one mile of the WSAR was finished that year.

Slides would take out other sections of the WSAR the next year, frustrating the crew's attempts to reach a stand of mature trees just past Seventynine Creek. Getting to this timber became more critical as road-building costs escalated. Each successive stretch of road that was built required more timber to be sold to pay for it. The immature forest found along the first few miles of the WSAR had been too young to harvest. The costs of the first few miles were covered almost entirely from timber sales on the east side of the tree farm.

The biggest engineering challenge along the WSAR was bridging Seventynine Creek. The creek starts near the divide between the Alouette and Stave Lake drainages, dropping almost eight hundred metres over several kilometres before entering the Stave north of Devils Lake. There was no "good" crossing point. Below the current bridge, the creek carves a deep channel through granite, before levelling out into a wide, braided channel.

In July 1966, crews began erecting a permanent fifteen-metre-long bridge more than thirty metres above the creek. Four old-growth Douglas firs, harvested just north of the crossing and each measuring at least one metre in diameter, were used for the decking. Crews anchored the bridge supports into the bedrock on the sides of the channel. As the crossing neared completion, the District auctioned off 210,000 board feet (495 cubic metres) of timber in the vicinity of Seventynine Creek.

Soon, crews moved into forest beyond the point that Japanese immigrants, working as bolt cutters in the 1910s, had reached. And still they had not even reached the halfway point between the tree farm's southern and northern boundaries. Nor did the job get any easier. Slides repeatedly

During the tree farm's first decade, Mission spent a quarter of a million dollars on building sixteen kilometres of road.

washed out parts of the WSAR. Crews spent much of 1967 drilling through granite before they could lay down the subgrade for the road. The bulldozers negotiated steeper terrain, and as the road climbed in elevation, winter weather became an issue. Snow and ice arrived early and stayed late, keeping crews off the upper WSAR until summer. Costs escalated.

During the tree farm's first decade, Mission spent a quarter of a million dollars (nearly two million in 2020 dollars) on building sixteen kilometres of road. Most of the expense went to just six kilometres of the WSAR. A decade later, the road was largely complete.

Once built, roads need regrading and regular repairs. Standards for road construction increase in both complexity and cost over time, and remediation of impacts on fish-bearing streams is required. Bridges and culverts get replaced. Logging roads may be deactivated. New roads are built along better grades or in less sensitive habitat, or to access new land acquisitions or maturing timber.

Today, the WSAR/FLFSR is the province's busiest forest service road. Tens of thousands of visitors annually flock to the shores of Stave, Devils, and Sayres Lakes. Increasing traffic compounds the difficulties and expense of maintaining the road. The WSAR still meets its original purpose—supporting the tree farm's logging operations—but more of the drivers on the road are travelling in cars better suited for city driving than for logging roads.

7

BRANCHING OUT

BY THE END of the 1960s, the District was farming tens of thousands of new Douglas fir trees each year. Fir provided for most of the tree farm's overhead. It paid for the roads and crews, the planting and equipment, and a small silviculture nursery. Fir commanded a much higher price than western red cedar and hemlock. New plantings would take more than seventy years to mature, but the wood's versatility made it a safe investment.

The relative stability of wood markets and growth in tree farm operations gave Rocky Rockwell the budget to hire Howard "Smokey" Murdoch as foreman. Rockwell and Murdoch expanded the scope of activities taking place on the tree farm, including research projects in partnership

PRECEDING A float trip up the Stave Reservoir in August. *Jason Brawn*

FACING Rocky Rockwell looks on as Minister Ray Williston plants the one millionth tree, 1971. *Mission Community Archives 0305-17*

with federal and provincial agencies. More time was spent on outreach and engagement with school groups, forestry students, and other municipalities. The demand for recreational amenities escalated. Crime did as well.

In short, as time marched on, the tree farm looked more like a community's forest, with all the problems and opportunities this status brings. Mission's commitment to local management would be tested during the 1980s, as a global recession bore down upon the timber industry. How the tree farm survived these rough years speaks to the strength of what had been built in the previous few decades.

For decades, Mission Forestry trained its own crews of students and seasonal workers to plant trees. Today, planting is contracted out. *Mission Community Archives 0305-18*

Planters

IN THE SIX decades between 1958 and 2018, planters spiked their spades into Mission's uneven, woodchip-littered forest floor more than 4.5 million times. The planters pulled back enough soil to make at least that many holes. Steel-toed boots struck the ground repeatedly to tamp the soil down around the 4.5 million seedlings that had been put there to rebuild a forest.

Almost every part of the forest industry is heavily automated, right down to the extraction of seeds from fir cones. Except this stage. Tree planting relies on back-breaking labour. While climbing hills, scrambling over stumps and pits, and identifying good spots to dig their next holes, planters carry heavy canvas satchels laden with young seedlings.

With respect to those still planting into their forties and fifties, tree planting is a young person's job. Planters work in all sorts of conditions and endure spring insect eruptions, foul weather, calluses, and sunburns. Unlike in most of BC, planters employed by the Mission Tree Farm could at least go home at night to comfortable beds and warm meals.

• • •

In the spring of 1960, crews planted more than thirty-five thousand trees. And this was a relatively inexperienced group of workers. Rockwell, the tree farm's manager, preferred it this way. He emphasized quality over quantity and deliberately hired eager but inexperienced tree planters whom he could train himself.

The next year, 55,500 more trees were planted. In 1962, the total more than doubled, to 119,000 trees. By the late 1970s, over 170,000 new trees were planted each year. The numbers fluctuated, based on how much cutting took place the year before.

While he was still in high school, Rockwell's son Lorne was hired as a planter on the tree farm, along with three or four university students and a regular complement of seasonal hires. Age worked to Lorne's advantage. "I got to be a very good packer and very good tree planter," he recalled. "I'd plant between eight hundred and a thousand. It wasn't that tough... I was twenty; these chaps were around forty, forty-five-ish.

"My dad used to say, 'I don't care how many trees you plant, just plant every one really well. Make sure every one you plant grows. It doesn't matter how many you plant, just make sure you plant them properly, correctly.' But being a competitive jock kind of guy, I just wanted to plant more, more, more. And they tested my trees. They'd come along and try and pull them out. That was the test. If you could pull a tree without breaking the top off, it wasn't planted properly."

Lorne's skills and speed as a tree planter landed him a permanent job with the tree farm. In the woods, he took his cues from Alex "Pike" Cameron, one of Rockwell's long-time crew. Cameron was a lifelong resident of Steelhead, born to one of the community's pioneering families. (Near Steelhead, Pike's Pond and Doreen's Trail are named after Pike Cameron and his wife.) Cameron already had decades of experience in the industry by the time Lorne came to work with him. In the summer when the conditions were too warm and dry for tree planting, Cameron would bring Lorne and the other planters up with him to work on salvage, splitting and packing cedar shakes.

After a few years, Lorne was taking on more complex responsibilities, operating bulldozers, compressors, and oil rigs and carrying out salvage operations. And he was getting older. When, in 1979, Murdoch asked him to go back to tree planting, Lorne said no. He left the tree farm to go work with a small crew working cedar salvage along Norrish Creek east of Mission and left tree planting to the younger crews.

• • •

Up until the late 1960s, almost all of the seedlings planted on the tree farm—90 percent of which were Douglas fir—originated in Surrey, at the BC Forest Service's Green Timbers Nursery. The site today is surrounded by high-rise residential towers and busy highways, but for decades, Green Timbers supplied seedlings to forest operators across the province. Starting in 1965, Mission supplemented the BCFS-supplied seedlings with trees grown in its own nursery.

The heavily reliance on Douglas fir had its drawbacks. At high elevations, heavy snows snapped limbs and smothered young trees. At lower elevations, rabbits and field mice dined on the seedlings, especially near Devils Lake. To improve the odds of longer-term survival, municipal crews returned to planting sites each spring and fall. Equipped with portable sprayers hoisted on their backs, they applied spot treatments of fertilizers and herbicides to stop elderberry, salmonberry, and other brushy species from shading out the young trees. The District also hired seasonal workers to pull weeds.

As the trees aged, crews trimmed limbs and thinned the forest by removing excess tree growth. The elder Rockwell spent many a solitary afternoon pruning the trees himself. "Rocky Rockwell wasn't really an office guy. He used to go out into the plantation and prune the trees," remembered Graham Webster, whose father Bruce served as tree farm manager after Rocky.

Lorne Rockwell recalled, "My dad used to get in the office in the morning, do all his paperwork, and very seldom did he not go out in the bush in the afternoon. What he did in the bush was prune trees in the plantation... That was his passion; he'd prune acres and acres and acres of timber."

In 1971, Mission commemorated the one-millionth tree planted. On a damp, cool spring day, a small audience of local officials, press, and

forestry crew members gathered at the District's silviculture nursery. There, they looked on as the minister of forests, Ray Williston, planted—what else—a Douglas fir.

• • •

The District was adding between 130,000 and 160,000 trees each year. By 1977, it had surpassed the two million mark. Most planting was carried out in the spring, by both year-round and seasonal crews.

Fir still dominated, but Rockwell and Murdoch ordered more western red cedar and balsam from the Forest Service's nursery and later from a private supplier. Crews transplanted yellow cedar seedlings grown in Mission's own nursery at higher elevations where they were better adapted than fir to the deep snows.

Ominous times loomed on the horizon. Changes to the tree farm licence, coupled with a deep recession, halted the momentum Mission enjoyed in the 1970s. Prices for timber plummeted. Because planting volumes are driven by harvesting numbers, planting slowed to a fraction of what was accomplished only a few years earlier. By 1982, fewer than fifty-two thousand trees were planted. Only thirty-five thousand trees went in the ground in 1983. The following year, fewer than twenty-four thousand were planted. No trees were planted at all in 1985.

By the late 1980s, Mission had emerged from the economic doldrums and was back to planting more than one hundred thousand trees a year. When the industry recovered, however, the business model for planting had changed. This included contracting out some of the work previously done in-house. Contracting drove down overhead costs and the cost per tree for planting. The same types of planters are employed, but they are working for companies with multi-year spring planting contracts.

After the recession, Mission also relied less heavily on Douglas fir. Today, planters heading up into the hills for a day's work on the Mission Tree Farm are slinging satchels weighed down with western red cedar seedlings.

FACING The Devils Lake loop trail winds through mature Douglas fir forest. *Michelle Rhodes*

Cultivators

FOR CENTURIES, THE Japanese have prized hinoki (*Chamaecyparis obtusa*) for its durability, strength, fragrance, and golden colour. The slow-growing cedar tree, native to Japan, is resistant to decay. Japanese builders used old-growth hinoki wood for important buildings and artifacts, and hinoki logs were incorporated into Shinto temple shrines and temples.

Japanese forests were heavily depleted during World War II, and most of the remaining old-growth forests disappeared from the countryside not long afterwards. Mature hinoki became scarce. The Japanese went in search of substitutes.

The preferred alternative, yellow cedar, grew on the Mission Tree Farm. The yellow cedar is a distant relative of hinoki and shares many of the same qualities, including tint and tensile strength. Both species are slow-growing. The Stó:lō value yellow cedar in much the same way that the Japanese did hinoki, using it for totems, longhouses, and boxes.

Starting around 1970, Japanese importers sought out yellow cedar supplies up and down the Northwest Coast. The greatest volumes were harvested in Alaska. BC producers tapped into the market as well.

Only a few years before, tree farm manager Rocky Rockwell and foreman Howard Murdoch had identified tracts of old-growth, high-elevation balsam, hemlock, and red and yellow cedar to be auctioned for harvest. The timber had been inaccessible prior to the completion of the West Stave Access Road. Many of the trees were massive, some upwards of ten feet in diameter and some nearly seven hundred years old.

Rockwell and Murdoch's interest in yellow cedar extended into developing new silviculture research trials. They spearheaded the first attempts in Canada to reproduce yellow cedar seedlings from seed. And they did so in the District's own nursery.

• • •

The nursery, after it was built in 1965, served as the nerve centre for the Mission Tree Farm. The facility sat on a small parcel of land near Steelhead. Rocky Rockwell, Pike Cameron, Howard Murdoch, and later Howard's brother Pat, Rocky's son Lorne, and other forestry crew members would rendezvous each morning at the nursery.

The initial rationale for the facility was to improve the survival rates for the two-year-old seedlings purchased from the BC Forest Service. Mission's crews transplanted the seedlings onto the nursery grounds to allow them to mature for another year or two. The first batch of ten thousand trees from the nursery, primarily fir, were ready for planting in late 1966.

The nursery also provided space to conduct experiments in silviculture. At first, these consisted of small projects, such as treating young trees with varying doses of herbicides to determine safe dosage rates. Later, the nursery was home to pioneering yellow cedar seed trials. The decision to begin working with yellow cedar followed from successive winters of heavy snows that had damaged young Douglas fir trees. Rockwell and Murdoch, wanting to switch to more resilient yellow cedar at high elevations, encountered a problem: the BC Forest Service did not supply yellow cedar seedlings.

Mission partnered with the BCFS in 1970 to begin yellow cedar cone collection and seed extraction. "Instead of waiting for someone else to put the idea into effect," recalled Murdoch in 1977, "we just went ahead and did it." The early partnership with the province would be the first of its kind in Canada.

Collectors gathered cones from yellow cedar, fir, and hemlock from the Mission Tree Farm, but it was yellow cedar that proved to be the most demanding. Reproducing yellow cedar requires "activating" seeds out of their dormancy, in a process referred to as stratification. Even then, the seeds may not prove viable. No one had stratified yellow cedar seed before, and processes that worked for other species did not necessarily apply. The high-elevation habitats and climatic conditions preferred by yellow cedar were also difficult to replicate in a nursery setting.

The BCFS's Green Timbers staff took on the initial testing and manipulation of seeds from Mission's tree farm. The seeds were also shipped to the BCFS's experimental nursery in Duncan, on Vancouver Island. Agency scientists were unsuccessful in their attempts to establish yellow cedar starts.

Frustrated by the lack of progress, the District asked the Federal Research Department in 1971 to get involved. Half of the seeds collected the next year were sent to the federal facility in Victoria, and Mission held on to the rest. With the support of Mission council, whose members were tantalized by the high prices yellow cedar would fetch, Rockwell's team pursued its own stratification work, in consultation with but independently of the federal research group.

Up until this point, most of the work took place in sheds or in fields on the nursery grounds, but the yellow cedar trials demanded more space. A greenhouse was installed, and a dedicated attendant was hired in 1974. The greenhouse allowed nursery workers to start seedlings under controlled conditions, rather than relying on one- and two-year-old transplants acquired from elsewhere.

Getting yellow cedar seeds to germinate and take root took a fair amount of trial and error. Seed stratification involves recreating the natural conditions in which a species would normally thrive. Yellow cedar prefers colder environments, so to simulate winter-like conditions, nursery workers refrigerated the seed. Simply chilling yellow cedar seed, however, turned out to be insufficient to coax the seed out of dormancy.

Mission's tree farm crews figured out that yellow cedar required a more complicated regimen to trigger germination. After four months spent at freezing temperatures in the fridge, workers removed the seed and transplanted it into styroblock trays filled with a loose mixture of peat and gravel. After a comparably mild six weeks in the greenhouse, the seeds—still in their trays—were then re-winterized in the refrigerator for

The greenhouse allowed nursery workers to start seedlings under controlled conditions, **rather than relying on one- and two-year-old transplants acquired from elsewhere.**

Seedlings planted on the tree farm, like these Douglas firs, are grown in private nurseries outside Mission. *Kelly Cameron*

another season. The results were immediate. In 1975, Rockwell's team successfully grew more than twenty thousand yellow cedar plants, a number that doubled within the year.

Once established, yellow cedar and other species fared better indoors than outdoors during their first year. Prior to the construction of the greenhouse, all of the planting had taken place on the nursery grounds but with limited success. "Last year we ended up with only 800 saplings from the 10,000 seeds we planted outside," Rockwell told the *Fraser Valley Record* in 1974. In contrast, the nearly constant 70°F indoor temperatures encouraged young tree growth at a consistent and predictable rate.

Mission's nursery crews experimented with mountain hemlock, grand fir, and amabilis fir (also known as Pacific silver fir). They continued sending the BCFS cones for continued trials. The tree farm's greenhouse filled to capacity, and the fields surrounding the building were planted with rows of two-year-old yellow cedars. Crews planted the first generation of yellow cedar progeny across more than one hundred acres of higher-elevation land in the tree farm. Where young Douglas fir struggled under deep snows, yellow cedar survived and thrived. By the end of the decade, yellow cedar survival rates on the Mission Tree Farm topped 95 percent.

• • •

After Rockwell retired and Murdoch moved on from the tree farm in the late 1970s, the work at the nursery continued for a few years more. Graeme Hynd took over as tree farm manager, succeeded the next year by Bruce Webster. Both were committed to continuing Mission's silviculture efforts.

Budget pressures forced a series of difficult decisions throughout the 1980s. As new cutting diminished to the minimum required for the licence, the need for seedlings dropped. Mission could not grow its own trees at a lower cost than the BCFS and private suppliers. Experiments in silviculture became a luxury that the tree farm could ill afford in the midst of a recession. Webster had little choice but to close the nursery for good in the mid-1980s.

At some point, a nursery may return to the tree farm, and yellow cedar may be the reason why. Climate change threatens to dramatically reduce the tree's range, possibly erasing it from the Mission forest altogether by mid-century. A field nursery would provide the space needed to assess whether more resilient variations of yellow cedar and other native species could survive in the Stave watershed.

Today, the site of the original tree farm nursery remains unused and unrecognizable. The road accessing the site is unmarked. The greenhouse and sheds are gone. A maturing canopy of trees, many of which started as outplantings, spreads unbroken across the old nursery grounds. And each morning, crews check in at the department's offices in the municipal hall before heading out to the woods for the day.

Researchers

FROM THE RIGHT-HAND side of the helicopter, Howard Murdoch watched the fire activity down below. He and the pilot were flying over Stave Lake, just north of the dam. Howard leaned forward and positioned his camera against the glass door. He was documenting the fire from the sky.

An hour earlier, someone on Murdoch's team—a graduate student, maybe, or a forest technician—had lit a diesel fuel torch and set flame against wood waste. Piles of broken treetops, branches, and stumps smouldered, then produced flames.

How fast did the fire move? Did the fire shift in the direction that was anticipated?

Sunny, cool, crisp days like the one spent in the helicopter provided optimal conditions for tracking fire behaviour. Columns of white and grey smoke built upwards and mushroomed. The plumes were visible across the central Fraser Valley. On the ground, forest scientists, technicians, and university students monitored burn conditions.

How much of the waste burned? Did the fire take out healthy trees? Were conditions too dry? Could the fires be managed?

Howard "Smokey" Murdoch came by his nickname honestly. He loved studying the dynamics of fire and advocated using fire as a means of managing logging waste. He conducted experiments on using fire as a management tool. Murdoch taught foresters across BC how to use fire safely and effectively, and he used the Mission Tree Farm for some of these trials.

Murdoch, though, was a fire mechanic more than a fire ecologist. Understanding the impact of fire on the health of forest ecosystems requires detailed, longitudinal studies. These would have to be carried out by government- and university-affiliated fire scientists. Both Murdoch and Rockwell were keen to see the tree farm used for scientific research, which they viewed as keeping Mission's commitment to progressive forest management. The first major research program to take place on the tree farm, a five-year study of fire's effects on vegetation, was announced in July 1968.

• • •

In the 1960s, North American forestry was less than a century old. At the University of British Columbia, the province's flagship university, the focus was on teaching intensive forestry and its operational needs. Students learned how to design and put in roads, culverts, and bridges, how to estimate cutting volumes and to identify replanting needs. Forestry school graduates learned less than their peers today would on a forest's ecological process, fire behaviour, insect infestations, natural reseeding, or management for multiple uses.

UBC operated a small research forest just west of Mission. The university obtained the land in the early 1940s at the urging of its faculty, and of one person in particular—F. Malcolm Knapp. Thousands of forestry students received field training at this location. Before coming on in 1980 to oversee the Mission Tree Farm, Bruce Webster managed UBC's research forest from 1965 to the mid-1970s.

The Canadian Forest Service (CFS) also invested heavily in applied forestry research during the 1960s, expanding field sites and employing teams of researchers. One of these was Robert Henderson, a twenty-five-year-old fire scientist and recent graduate from the University of Montana's forestry program. Henderson joined the Pacific Forest Research Centre (PFRC) as a research officer in 1967. He then brought his friend Randy Lafferty, a fellow Montana alumnus and fire ecologist, north to join his team.

Henderson and Lafferty were interested in using fire to improve natural regeneration. Their goal was to determine how long and how hot fires should burn, and over how wide an area, in order to achieve the desired ecological effects. They faced significant resistance to conducting field trials, however. The project required finding a tree farm willing to interrupt operations and take on the liability of controlled burning. Unsurprisingly, most licensees turned them down.

TFL 26 was the only tree farm licence not in private hands, and Mission surfaced as the only volunteer willing to partner on the study. After "much discussion," according to Rockwell, the CFS and the municipality struck a deal in 1968. Years later, Rockwell reflected on his decision to support fire experiments on the tree farm. "Nobody was willing to take the responsibility of having a fire get away on them," he said. "You have to take chances."

Nonetheless, Mission's tree farm was less than ideal for these experiments. In an interim report for the PFRC, Lafferty identified the tenure's "broken topography, small logging operations, diversity of logging areas and wet climate" as challenges to carrying out the research. Burning also took place in cutblocks that the research team had little say in selecting. On the other hand, Mission's milder climate permitted controlled burns over more of the year, including during summer. That would be unthinkable in BC's Interior.

The project team carried out twenty-two burns between July 1968 and September 1970, splitting their time between higher-elevation forests on Mt. Crickmer (elev. 1,356 metres), and lower-elevation forests on the tree farm's eastern boundary. Henderson and Lafferty were joined by meteorologist Jack Turner, crew chief Keith King, a technician, and six university students. The tree farm hired local students to help out as assistants. American researchers from the US Forest Service and the University of Washington came up to take measurements of air pollution levels resulting from fire activity.

The data collected on fire behaviour during the controlled burns would be used to help develop fire application guidelines. For his part, Lafferty continued collecting data on vegetation in burned areas for the next decade, examining the relationship between fire intensity and location and plant regeneration. His work on the project continued even after he left the CFS to work in the private sector. In the mid-1980s, a soil scientist at UBC returned to some of these sites, where he conducted new burning and soil testing.

Graduate students and other researchers watch from a safe distance as deliberately set fires consume slash left over from logging. *Mission Community Archives 0305-19*

The Mission Tree Farm also played a much smaller role in later ecological research. This research has largely involved the collection of biological material taken from the tree farm and analyzed in a university laboratory. In 1974, for instance, a pair of biologists from Simon Fraser University drove east to Mission to collect the budding ends of Sitka spruce branches, as part of a study to determine how quickly white pine weevils (*Pissodes strobi*) develop a taste for the spruce. (Quite quickly, as it turns out.) Longitudinal research occurs more frequently at UBC's forest in Maple Ridge, where logging activity is less likely to disrupt study sites.

• • •

Over the past two decades, most of the research conducted on the tree farm has focused on the area's archaeological and anthropological record. This research has reinforced and added greater definition to the overlapping Indigenous histories of the Stave watershed. In a series of studies first commissioned by the Kwantlen First Nation, the Stó:lō First Nation, BC Hydro, and other agencies, archaeologists began systematically recording the history of the Stave River.

Much of this work was carried out by anthropologist Duncan McLaren, who completed his master's thesis research on the history and archaeology of the Stave watershed. Initially, the commissioned research was intended to inform water use planning and subsequent upgrades to dam facilities. The archaeological record of the Stave region has expanded as other researchers—many of whom work with McLaren—have completed detailed studies. Most of these reports are not publicly available, because in some cases circulating the reports would put vulnerable sites at risk.

None of these studies would have been possible or credible without the involvement and support of the Kwantlen and Stó:lō First Nations. Archaeological investigations are now regularly carried out as a condition of cutting activity in Mission's forest. In the years to come, the expectation is that most research, archaeological or otherwise, will entail continued engagement with the Matsqui, Leq'á:mel, Katzie, and Kwantlen Peoples.

Fire ecologist Robert Lafferty of the Canadian Forest Service assesses site conditions for new seedlings. *Mission Community Archives 0305-20*

Vandals

OVER THE YEARS, at least some Mission residents have interpreted "tree farm" quite literally. Each December, armed with hacksaws and tie-down ropes, fir hunters have driven up to the forest to find that perfect Christmas tree.

Each pilfered Douglas or grand fir adorned in lights, baubles, and garlands standing sentinel in a picture window costs the tree farm money. Poachers typically seek out younger trees of ten to fifteen years of age. These are plantation firs that the District paid crews to transplant. Left to grow to maturity, a fir tree would be worth tens of thousands of dollars. Abandoned curbside after the holidays, strands of tinsel wrapped inextricably around its branches, the tree is worthless.

For years, the District hired patrols over the weeks and nights before Christmas to stop cutters in their tracks. Limited road access west of Stave Lake initially deterred poachers from the area, but there was plenty of opportunity near Steelhead. Snow helped patrols track tree hunters and firewood collectors, but it often did not arrive until later in the season. At best, the patrols were a chance to educate locals. More realistically, if someone really wanted a tree, there was little in the way of them getting one.

Mission's community forest operates close to larger and growing urban areas, and this invites any number of challenges. Christmas tree poaching has been costly, but it's tame compared to some of the other problems tree farm managers have faced over the years. Vandalism. Log theft. Dumping. Shooting. Vehicle break-ins. Drunk driving and vehicle accidents. And sometimes, worse.

Thinly staffed tree farm crews could not tackle all problems simultaneously, and over time, they ended up spending more on repairing shot-up and damaged equipment and signage. In the 1960s and 1970s, the District budgeted for vandalism as part of its forestry budget. The costs of the damage caused by vandals and thieves were not necessarily unmanageable—at least not in the first couple of decades. It was that criminal activity created new risks for crew, contractors, and the public. And both the risks and the costs escalated as area populations grew.

These problems vexed Rocky Rockwell and each subsequent tree farm manager. Not long after Bruce Webster took over, he instructed crews to install a gate on Hoover Lake Road. The decision came down after thieves stole high-value cedar logs and others engaged in uncontrolled firewood cutting along the road. The move signalled the beginning of the end of uncontrolled vehicle access on the eastern half of the tree farm.

By the 1980s, Mission's Forestry Department allocated thousands of dollars each year to deal with vandalism, including to logging equipment. Thousands more went to pay for workers on summertime fire patrols who snuffed out unattended and unpermitted campfires. The time spent on clean-up and repairs took away from other priorities, including public education.

The worst offences took place west of Stave Lake. After the completion of the West Stave Access Road, municipal crews laid out secondary logging roads up to higher elevations. The public had access by car and ATV to more of the tree farm than ever before. This was great for those wanting a secluded camping spot along a peaceful lakeside location. It also worked well for those who, in 1974, set up a marijuana grow op "a long ways from nowhere" so as to not be detected.

The western half of the forest attracted target shooters who, in small numbers, might once have been tolerated, even though shooting was prohibited. But the sound of gunfire was frequent and piercing, and the

Christmas tree poaching has been costly, but it's tame compared to some of the other problems tree farm managers have faced over the years. Vandalism. Log theft. Dumping. Shooting. Vehicle break-ins. Drunk driving and vehicle accidents. And sometimes, worse.

damage done not insignificant. Shooters peppered road signs, trees, and even logging equipment with buckshot and bullets. Most frustratingly, many of these same folks would have driven past the established (and perfectly legal) Mission Rod and Gun Club target range on their way to the woods.

Most who ventured further into the tree farm did so for legitimate reasons—to camp or fish. But the situation along Burma Road devolved further during the 1990s. The number and diversity of criminal activities only increased. By the early 2000s, the western half of the tree farm had earned a reputation—a "wild west" where seemingly anything goes. Cleaning up the area became the biggest priority and most difficult challenge yet for the Mission Tree Farm.

Inmates

TODAY, NINE PRISONS operate in the Fraser Valley, provincial and federal, for men and women, from minimum to maximum security. Decades ago, even more corrections facilities were in operation, including some that were found in Mission's forest.

The practice of housing inmates in forestry camps far from towns and families dates back to the end of World War II. Work camps sprang up from Maple Ridge to Chilliwack beginning in 1950. The provincial government viewed these facilities as cost-effective strategies for rehabilitation. Inmates left the camps having learned a trade in silviculture, logging, or milling.

In 1967, BC Corrections began developing camps specifically for young adult offenders. Out of this initiative came a dedicated forestry and wilderness survival program at Boulder Bay (Alouette Lake) in north Maple Ridge, based on the Outward Bound model. Candidates selected for the physically demanding program arrived at the camp from around BC. Each inmate spent four months living off grid in the self-sustaining facility. The Boulder Bay Camp was followed by three other specialized camps: Pine Ridge in Maple Ridge and camps at Stave Lake and Cedar (Sayres) Lake in Mission.

Corrections built the Stave Lake Camp in 1971 as part of a partnership with the Ministry of Forests and BC Hydro. Like Boulder Bay, the facility operated as a young adult offenders camp for those serving sentences of less than two years. Beginning in 1988, BC Corrections also transferred inmates convicted for first-time sex offences and serving short sentences to the Stave Lake Camp.

The Stave Lake Camp was located seven kilometres north of the dam along the West Stave Access (Burma)Road, on land excluded from (but surrounded by) TFL 26. The Ministry of Forests had previously set the land aside for future park development. Mission's forestry staff worked with Corrections to train inmates in the logging techniques needed to clear the site.

Once the grounds had been logged, Corrections crews began harvesting underwater stumps and logs near the shoreline. The massive stumps dated back to the area's early logging history, before the Stave Falls Dam was erected, before the lake rose and submerged the remnants of low-lying, cutover forest. The reservoir's chilled waters entombed and preserved the wood. But the stumps presented a hazard to boaters and other users and an impediment to park development. Over several summers, BC Hydro drew down the reservoir long enough for inmates to do both logging and clean-up of the foreshore. By the end of 1976, they had cleared more than thirty acres.

During the winter months, camp crews converted the harvested stumps into firewood. So much wood was harvested that crews were able to supply campgrounds as far away as E.C. Manning Provincial Park, more than one hundred kilometres to the east. The Ministry of Forests trained inmates to fight fires, construct trails, and maintain Burma Road. Inmates were also employed in processing cedar shakes.

The camp even supported wildlife rehabilitation and hatchery operations in the mid to late 1990s. Inmates tended to injured birds that had been found poisoned or injured around the Lower Mainland. The Stave Lake Camp aviary housed peregrine falcons, golden eagles, great horned owls, and other birds of prey. Inmates were trained to care for the birds, maintain pens, and help the birds practise flight until they were well enough to be reintroduced into the wild. At Sayres Lake, inmates

The Stave Lake Camp aviary housed peregrine falcons, golden eagles, great horned owls, and other birds of prey. **Inmates were trained to care for the birds.**

operated a fish hatchery using floating pens anchored off the lake's south shore. The lake, too cold to support a diverse native fishery, sustained the production of hundreds of thousands of rainbow trout and other species. Fish collected from the pens were used to stock other lakes in the region.

BC Corrections also ran a second short-lived camp on the southwestern shore of Sayres Lake between 1976 and 1979. The Cedar Lake Camp housed repeat offenders with less than two years left on their sentences. When not in camp, inmate crews spent their time on the Mission Tree Farm, pruning and cutting trees and bundling cords of firewood.

Following its closure, the Cedar Lake Camp site was plagued by problems. Broken generators, road washouts, and structural problems with some of the buildings drove up the costs of maintaining and letting out the facility. A multitude of possible user groups came and went, renting the site temporarily or occasionally proposing long-term occupancy. The camp met its demise in the mid-1990s, after vandals torched the buildings.

Today, the same area is home to a forty-site campground. The District removed the fish pens after the hatchery closed. The floating walkway that led out to the pens collapsed years ago. Intact sections of the boardwalk and the greying timbers that once supported it drift along the southern shoreline, at the entrance to Sayres Creek.

The Stave Lake Camp operated for more than three decades. BC Corrections closed its camps in 2002, part of a broader reorganization in which nine facilities were shuttered. The agency left behind a fully landscaped and serviced park-like setting just south of Kearsley Creek, including bunkhouses, workshops, an industrial kitchen, and other buildings. In 2003, the Zajac Foundation acquired the facility and grounds. What was once a site for prisoner rehabilitation has been transformed into an adapted summer camp compound for children with special needs.

The partnership between the District of Mission and BC Corrections has not completely disappeared, however. When volunteers are willing and arrangements can be made, inmates from the Fraser Valley Regional Correctional Centre still help with trail maintenance and building. The prison camps, however, are gone, and likely for good.

WHITE
TRUCKS
TIM
CAT

Students

WHEN THE CEDAR Lake Camp shut down in 1979, the province transferred ownership of the nine-acre site to the District of Mission. It did so on the condition that the site be used for "environmental, education and community recreation purposes only."

In the decade that followed, multiple organizations used, or tried to use, the camp. At the time, the site consisted of little more than a handful of bunkhouses and trailers. Groups from as far away as Vancouver volunteered their time to clean up the location. Both the Mission and Maple Ridge school districts expressed interest in using the camp.

One of the groups that expressed interest in using the camp was the Sayres Lake Camp Society (SLCS). The District of Mission leased the whole site for one dollar to SLCS, whose membership included Mission Tree Farm foreman Howard Murdoch. The society had planned to run an independent environmental centre, that is, until severe rains damaged buildings and washed out the main road during the group's first year.

SLCS's vision to create a dedicated educational facility was the latest in a series of proposals that dated back to before the tree farm licence was even allocated. The Mission School Forest Association and the District had operated a dedicated forest plot in the 1950s and 1960s, where senior high school students completed courses and training in forestry. The closure of the Cedar Lake Camp provided the District and the tree farm with another opportunity. But the site's location—eleven kilometres up a rough gravel road—doomed it at the time. (The road has been improved since then.)

FACING Local kids build a fire while on an excursion to Sayres Lake. *Mission Community Archives 0025*

Forest education efforts continued despite the challenges of maintaining a dedicated educational site. The uniqueness of Mission's tenure arrangement and the tree farm's earlier yellow cedar nursery trials have attracted visitors from US, Guatemala, Colombia, Sweden, the former East Germany, and Italy. Faculty and forestry students from the BC Institute of Technology (BCIT), Selkirk College, and College of New Caledonia (Prince George) have made frequent, sometimes annual trips out to Mission for almost forty years. Classes from the nearby University of the Fraser Valley (UFV) carry out projects and field trips on the tree farm.

In the tree farm's more than six decades of operation, tens of thousands of K–12 students from nearby schools have toured it. Busloads of fifth-graders have walked the Stave Dam Forest Interpretation Trail since it was completed in 1994. Groups of younger schoolchildren hiking the Roy Kittles Trail have stopped to form human chains around a Douglas fir hundreds of years old. When classes could not make it out to the forest, staff have brought the forest to them, giving animated talks and passing around cones and tree cores and other visual aids.

The Mission Tree Farm staff even contracted with a filmmaker to produce a video for school use. *Our Forests, Our Future* features a mix of original footage from tree farm operations interspersed with local officials and forestry staff standing awkwardly in front of the camera and reading from cue cards. Completed in 1989, the District distributed copies of the video to local schools. Anyone wishing to view the remaining copies of the video would first have to locate their school's last working VCR.

Since 2014, the Mission Forestry Department and its community partners have pushed to expand educational opportunities through the dedication of outdoor learning spaces and the creation of an Outdoor Learning Alliance (OLA). The OLA is an evolving network designed to connect educational organizations with an interest in using the tree farm for classroom and research activities, ranging from hour-long field trips to multi-day training camps. As of 2019, Mission public schools, BCIT, UFV, and Riverside College have signed on. The OLA, like the MSFA and the Sayres Lake Camp Society, may well advocate for dedicated research sites and facilities on the tree farm.

Technologist Kelly Cameron visits with a class at Stave Falls Elementary in 2021 to demonstrate how to identify trees in the forest. *Chris Gruenwald*

Crisis Managers

IN THE FIRST six decades, only six people oversaw the Mission Tree Farm. From 1958 to 1978, Rocky Rockwell prioritized investments in road building, planting, and silviculture. From 1989 to 2019, Kim Allan and Bob O'Neal worked with an experienced forestry team to modernize operations and pursue a progressive vision of community forest management with expanded public education and reconciliation.

The decade in between—the 1980s—belonged largely to Bruce Webster. It was his job to keep BC's first community forest initiative alive in a time of crisis.

• • •

Rockwell's retirement in 1978 was followed by the departure of the tree farm's foreman and Rockwell's right-hand man, Howard Murdoch, a year later. The District named Graeme Hynd as Rockwell's replacement. Hynd's time in the role was brief and marked by increasing uncertainty. His eighteen-month run saw good log prices and continued success with yellow cedar silviculture. The province, however, was preparing to implement the recommendations of yet another royal commission, including one that would change how Mission paid for the trees it cut.

Chaired by Dr. Peter Pearse of UBC's School of Forestry, the commission investigated the effectiveness of the existing tenure system, current patterns of how forest resources were being used, and the structure of the

Retired tree farm manager Bruce Webster (L) with the former reeve Neville Cox (R) in front of the vintage ambulance they restored together. Fraser Valley Record; *courtesy of Graham Webster*

forest industry. When the commission formed in 1976, a handful of large timber companies held the majority of BC's tree farm licences. Despite the province's stated aims of sustained yield, both the supply and quality of timber was declining fast. Expanded and spidering road networks contributed to what Pearse referred to as a "massive assault on the province's forests." Companies were moving into marginal timberlands, spending more to harvest timber that was less valuable.

TFL 26 was the only community-held licence of the thirty-four issued by the province; the others dwarfed Mission's in both size and timber volumes. The commission believed that the "success of the Tree-farm Licence held by the District of Mission, in the Fraser Valley, can be repeated elsewhere." Nonetheless, the policy recommendations the commission made threatened the tree farm's very existence.

Since first being issued TFL 26, the District had paid a statutory royalty for the right to harvest timber. Other licensees paid appraised stumpage, fees assessed based on the value of timber harvested. These fees varied based on market conditions, but stumpage was almost always higher than the royalties the District paid. Rockwell had been quite upfront about how the royalty-based arrangement afforded Mission the financial buffer it needed in those first decades.

The commission argued, however, that the use of royalties rather than stumpage constituted "a rather arbitrary transfer of revenue from the province as a whole to selected municipalities," that is, to Mission, "and has no apparent logic in either forest management policy or municipal finance." As the BC Legislature moved to implement the recommended changes in 1979, Hynd and Mission alderman John Parkes travelled back and forth to Victoria to lobby for a continued exemption. A frustrated Parkes told a reporter for the *Vancouver Sun*, "We don't think it's fair that we should be treated the same way as the other tree farms which cover millions of acres and are held by large corporations."

The impacts of having to pay appraisal stumpage were immediate and drastic. Mission paid out more than $200,000 in combined royalties and stumpage for 1979 alone. The next year brought changes to how the tree farm charged harvesters for timber. The District could only afford to cut a third as much wood as it had the year before. In the midst of this difficult transition, Hynd and the District parted company.

• • •

Bruce Webster had unsuccessfully courted the job of tree farm manager when Rockwell retired. Webster had wanted to stay in Mission, where he raised his family, coached soccer, served on the hospital board, and chaired the municipal recreation committee. He knew the tree farm's crew the way that anyone working in the same industry in the same small town would. His kids went to school with their kids.

Like Rockwell, Webster spent decades in the industry before becoming manager. The two men shared a friendship as well as Scottish roots and a love for silviculture. Born in Alberta in 1925, Webster and his parents moved to Scotland during the Depression. His first forestry job involved supervising Italian POWs planting trees. After four years as a pilot for the Royal Air Force, Webster signed up for a two-year forestry training program based out of Peebles, Scotland.

Webster followed his sister and her husband to BC in 1947. After he landed a job with the BC Forest Service, the agency sent him to ranger school at Green Timbers in Surrey. The BCFS transferred Webster to the Mission Forest District in 1959. From the corner of his living room, where he had set up a card table, Webster spent his nights completing the courses he needed to receive his registered professional forester status.

In 1965, Webster took on managing the UBC research forest in Maple Ridge. UBC's forest had a lot in common with Mission's tree farm, including its location between Alouette and Stave Lakes. Both tenures were experiments when they were established in the late 1940s—one as a research forest, the other as a forest reserve—and both were established on lands that had been heavily logged decades earlier. For nearly a decade, Webster oversaw the practical needs of the research forest: planning roads, building bridges, selling logs.

When the tree farm manager position opened up in Mission for a second time in less than two years, Webster was working as a private contractor near Harrison Lake. Despite having been previously passed over for the job, and knowing the precarious financial position the tree farm was now facing, Webster did not hesitate to take the job.

• • •

The significantly higher stumpage fees Mission had to pay came into effect just before a collapse in the market for wood products. Unemployment spiked. Trade battles with the US flared up. A deep recession set in. "You could hardly give away good timber," noted Graham Webster, Bruce's son, also a professional forester.

Mission had experienced some rough patches in the 1970s, but those were short-lived and less severe. Bruce Webster managed to find creative solutions to keep the tree farm afloat through a more prolonged slump. He kept the permanent year-round forestry crew lean. Webster relied on federal and provincial employment and forest development grants to pay for his seasonal crews. Most years, non-essential projects were shelved.

Mission also temporarily halted timber sales—not many people were buying anyway, and those who were wanted the timber for too low a cost. Webster's team instead contracted harvesting out directly to local loggers. This allowed Webster to take advantage of the occasional upticks in the market to get better prices for the wood. Some years, however, those opportunities never materialized.

The tree farm recorded losses in six out of the eight years between 1978 and 1985. Webster still had to find a way to meet the five-year average AAC, which became harder to do as logs cost almost as much to cut down as they earned in the marketplace. Buyers could afford to be more selective, demanding better grades of higher-diameter logs. Few buyers wanted smaller trees.

The political pressures from within the community mounted. "There was a councillor on Mission Council at that time, that proposed that if the tree farm couldn't make money that it should be turned into a trailer park. She thought that the trailer park would make some revenue," recalled Webster. "It was very frustrating for my dad... Very few people could understand the tenure system."

One of the recession's casualties was the nursery and greenhouse. The decision was difficult for Webster and not just because of his close friendship with Rockwell. Webster's Scottish training emphasized silviculture over forest engineering. "I think he really saw the significance of what was done there," Graham recalled, "but he was practical enough at that time to accept it."

The downturn came to an end, ironically, following a forest shutdown. In 1986, Jack Munro and the International Woodworkers of America (IWA), the union representing BC loggers, called a strike that lasted nearly five months. The shutdown cost BC more than $2.5 billion and halted the

Bruce Webster learned to manage a tenure at UBC's Malcolm Knapp Research Forest in Maple Ridge. Webster (bottom left) is pictured with Knapp (bottom right). *Courtesy of Graham Webster*

The impacts of having to pay appraisal stumpage were immediate and drastic. Mission paid out more than $200,000 in combined royalties and stumpage for 1979 alone.

flow of most newly harvested logs into the marketplace. As a result, log prices spiked. This coincided with a surge in activity on the tree farm, as Mission tried to make up for missing AAC targets the previous few years. The Mission Tree Farm raked in surpluses greater than the combined deficits of the previous five years.

Escalating revenues in 1987 and 1988 further put the tree farm back into the black. Webster tackled the surfeit of work put on hold during the recession by using log revenues and Section 88 project funding from the Ministry of Forests. He found additional ways to save money by using a single contractor for logging and booming and by selling logs to only one wholesaler rather than on the open market. When the province changed how stumpage was calculated, he found ways to absorb the increased costs.

Webster also created a "rainy day" fund. Forestry is a cyclical industry, and the swings were becoming wider and more frequent. To protect itself from deep losses, Mission established a Forestry Reserve account. A second account, the Income Averaging Reserve account, would be used to level out revenues when log prices dropped.

The funds created new political tensions. In the 1970s, the District's general coffers swelled with surpluses generated by the tree farm. The new accounts required that Forestry claw back some of these funds. Webster first did so in 1986, when he redirected a quarter of a million dollars into the Forestry Reserve. Soon, tree farm revenues were channelled into both accounts.

• • •

Webster retired from the tree farm in 1989, a couple of months after his sixty-fifth birthday. The District had already hired his successor, Kim Allan, and Webster felt some pressure to pass over the reins. A few years earlier, his wife retired from her job as a schoolteacher. Webster kept busy volunteering in town and restoring an old ambulance. Eventually, the Websters relocated to Campbell River but visited Mission often, including for the tree farm's fiftieth anniversary in 1998. Bruce Webster travelled for the festivities and helped cut the cake.

8 A COMMUNITY'S FOREST

FOLLOWING AN INDUSTRY-WIDE crisis in the 1980s, the tree farm entered its fourth decade on more stable financial footing. Mission set an example to the rest of the province of how resilient locally managed forests could be. But the crisis served as a reminder of the need to nurture political support within the community as well.

Kim Allan, appointed as Mission's first director of forest management in 1989, called for a renewed emphasis on communicating the forest's benefits to the town. He also called for a new name—the Mission Municipal Forest. Trail-building and public education efforts accelerated.

PRECEDING Returning from a morning hike on Reservoir Trail. *Jason Brawn*

FACING Forest technologist Elske von Hardenberg conducting a waste and residue survey after harvesting. *Kelly Cameron*

Under Allan and forestry manager Bob O'Neal, the tree farm evolved into the community's forest. This also meant a reckoning of sorts, one in which Mission would have to evaluate its relationships (or lack thereof) with its Indigenous neighbours. The tree farm operated on Crown and private land for which no treaty has been signed. Traditional uses long went unrecognized and unaccommodated. It was time to take the first steps towards reconciliation.

Past directors Bob O'Neal (L) and Kim Allan (R), and current director Chris Gruenwald (centre).
Chris Gruenwald

The Community Foresters

KIM ALLAN JOINED the forest in the role of the director of forest management in 1989. The title was new, although his responsibilities were much the same as Bruce Webster's had been. The following year, Bob O'Neal was hired as forestry manager. The director-manager system greater increased Mission's capacity to pursue projects on TFL 26. The Forestry Department also enjoyed low staff turnover and highly experienced crews during Allan and O'Neal's years with the tree farm.

Unlike earlier tree farm managers, both men came to Mission having spent their careers in the private sector rather than in the provincial forestry branch. Allan hailed from North Vancouver, although he had spent much of his early career on Vancouver Island. After finishing his forestry degree at UBC, Allan worked as a timber cruiser with BC Forest Products, then as a regional operational forester for Fletcher Challenge Canada's logging division. The directorship brought Allan back to the Lower Mainland.

The Ontario-born O'Neal also graduated from UBC Forestry, and both he and Allan were registered professional foresters. O'Neal worked up and down the west coast, first as a chokesetter when he was eighteen and then, after university, in forest engineering and planning for one of BC's major timber firms, Interfor. He later moved into the sales division of MacMillan Bloedel, at the time BC's largest timber company.

In the aftermath of the lengthy forest recession of the 1980s, their experience in the private sector proved useful. Allan and O'Neal knew how thin the margins of production could be and understood the importance of being adaptable. Under Allan's leadership, the department implemented better or more innovative business practices, such as contractor performance evaluations and niche market sales.

Allan's role was decidedly political and financial, working closely with Mission's council and District staff. It was his signature attached to forestry plans, budgets, and annual reports. And it was Allan who was ultimately held accountable for forest operations.

O'Neal assumed responsibility for managing the day-to-day needs of the tree farm, from laying out cutblocks for harvesting to meeting with contractors and consultants. O'Neal grew to know the forest intimately. In his twenty-plus years as manager, he remarked, he "pretty well walked most of the area."

Allan and O'Neal complemented each other's skills and approaches. Neither adhered strictly to their job descriptions when flexibility and task-sharing produced better results. Allan's passion was silviculture. O'Neal built roads. Allan did not much mind office work. O'Neal was happy to shed the necktie for Gore-Tex. Of the two, Allan possessed the quieter temperament, one that put residents and elected officials at ease. O'Neal's optimism and gregariousness—whether he was leading a school group or a public meeting—were contagious.

Allan and O'Neal worked together to champion the values associated with community-based management. They partnered with local recreation groups and schools to help build trails and other amenities in the forest. One of Allan's first moves as director was to change the name of the tree farm to the Mission Municipal Forest. This was more than an exercise in rebranding. The new name reinforced the commitment to community control.

Allan retired from the tree farm in 2013, after almost twenty-four years with the District. The differences between how the Mission Municipal Forest operated at the beginning of his time with the tree farm and at the

For the sixtieth anniversary, the community was invited to plant a tree on this logged parcel in Steelhead. *Michelle Rhodes*

end were substantial. The awareness of the forest among District residents improved, in part through spending on high-profile community projects. Mission Forestry's finances remained relatively stable, no mean feat given continued volatility in the timber industry. The operation's environmental footprint improved, following the province's adoption of the Forest Practices Code (FPC) in 1995, even though Mission bore increased costs associated the code's expanded planning and reporting requirements.

In the mid-1990s, log prices once again plummeted. Meanwhile, road costs shot up. Aging bridges and culverts needed replacement and higher traffic volumes necessitated more maintenance work on the Florence Lake Forest Service Road (formerly known as Burma Road). Some years, the tree farm operated in the red. Allan impressed upon the District council the importance of evaluating financial performance over multi-year periods. "It was important that over those five-year periods we were always in the black," he recalled, "and that we did fairly well."

Changing technologies improved the Forestry Department's ability to conduct inventories and plan roads. When Allan came in, digital mapping was agonizingly slow and expensive. By the time he left, anyone with a cell phone could use Google Earth to find a cutblock or a campsite on the TFL. On Allan's watch, the department knew more about what was growing on the tree farm and could better and more sustainably manage forest inventories, and the forest was made more accessible.

Following Allan's departure in 2013, O'Neal took over as director. In the six years he spent at the helm, O'Neal advanced projects first launched while Allan was with the department. The biggest single project involved comprehensive multiple-use planning for the western half of the TFL—an area now referred to simply as Stave West. O'Neal made it a priority to improve relationships with Indigenous neighbours and better facilitate management priorities shared by Mission, the Kwantlen First Nation, and other Nations.

• • •

Allan next entered into part-time retirement. He took some time to catch up on projects and do some travelling and then returned to consulting two to three days a week. To the east of Mission, Allan worked on contract for the Cascade Lower Canyon Community Forest, serving as its general manager for nearly two years.

Few if any foresters can claim to possess as much experience or knowledge of Canadian community forest management as Allan. Between Mission and Hope, Allan spent more than three decades managing local enterprises. As director, Allan wrote about Mission's experience in academic articles and served as a resource for other municipalities interested in developing their own community forests. The Mission Municipal Forest was an innovator and a leader in community forestry, but according to Allan, "we weren't trying to get our heads swelled with the idea."

He reflected, "I'd like to think we were an example and trying to do a good job."

Community Building

SOON AFTER TAKING over as director, Kim Allan recognized that investing timber revenues directly into community projects could generate goodwill among area residents and elected officials. "If you could take a million bucks and say, 'Here you go and it can be put into a fire hall or a library or some type of community amenity,'" recalled Allan, "it could be shown and seen that the municipal forest was contributing to the social part of the community, the safety part of the community."

He added that "money was going to get spent anyway" on these community needs. Forestry surpluses would be transferred into general revenues, and it would appear as though council was paying for projects through taxes. Instead, in this way, residents would know how logging in the municipal forest generated economic benefits for the town.

Mission's Forestry Department enjoyed healthy surpluses as timber markets rebounded in the late 1980s and early 1990s. This enabled it to fund some high-cost items in the District. In 1990–91, these included the construction of Firehall #3 in Silverdale and the fire truck that worked out of it. When the municipal recreation centre underwent expansion, forest revenues paid for the new ice rink and for upgrades several years later. Most visibly, revenues covered the entire $700,000 price tag for a new library and archives building in downtown Mission in 1994.

The Forestry Department maintained a reserve fund specifically for community arts and culture programs. These funds helped offset the

In addition to funding projects, the municipal forest supplied high-value timber to community projects and infrastructure.

costs associated with staging the Mission Folk Music Festival, for example. In total, the department spent more than half a million dollars over a decade on community arts and culture projects. Mission's council decided in 2002 to cover arts programs through general revenues, and Forestry discontinued its arts reserve account.

In addition to funding projects, the municipal forest supplied high-value timber and other forest products to community projects and infrastructure. Several of these can be seen today at Heritage Park in east Mission. Visitors congregate under a large picnic shelter built using logs harvested from the tree farm. Or they stop to have lunch at the restaurant in the park, located in a building erected using logs from the tree farm. As was the nearby washroom and kitchen facility.

Mission's reserve funds and targeted spending allayed concerns that the tree farm was a drain on municipal finances. Critically, this spending changed the perception that many Mission residents had of the forest. The tree farm, according to Allan, was about more than "just going down and cutting a bunch of trees."

Logs from the tree farm were used to construct Norma Kenney House and other buildings at Mission's Heritage Park. *Michelle Rhodes*

The Technologists

MUCH OF THE work that Kelly Cameron does to grow trees on TFL 26 happens on a pair of computer monitors, placed side by side and plugged into the computer tower resting under her desk. Cameron's office is immaculate, with enough space on her desk to spread out maps (which she made), printouts of figures (which she crunched), and reports (which she compiled).

Forestry is still a profession spent outdoors, but it also demands a tremendous amount of information and planning back at the office. Almost every piece of data on the tree farm—every tree-growth measurement, every order for seedlings, every point on a map—is recorded and analyzed digitally from there.

The nature of Cameron's work as the District's forest technologist requires precision and a fair amount of "ground truthing." She visits harvested cutblocks to measure the volumes of wood waste left behind by logging contractors. Cameron oversees the silviculture needs of the TFL. This means that she orders seedlings for the next year's planting and, back in the forest, estimates survival and growth rates.

Because she works for a community-managed operation, Cameron also handles many of the requests that come in from the public. She responds to emails and calls from local residents asking how planned logging will change the views from their properties. It is usually Cameron leading tours and giving presentations to elementary school kids.

The list of tasks she is responsible for varies from season to season, but if there's a question who is in the know, it's best to assume it is Cameron. She has spent two and a half decades working on the tree farm. She's mapped the same hills and watercourses many times over the years. On her days off, she hikes in the forest and takes photographs from her favourite spots.

• • •

In 1995, the Government of BC implemented the Forest Practices Code. The FPC emerged out of public concerns over insufficient protection for old-growth forests and ecosystem health. The legislation consolidated existing and new regulations and guidelines under one umbrella. The FPC introduced more stringent watershed protections, placed limits on the scale of logging activity on Crown land, and established a governing board to oversee enforcement.

Mission was not exempt from the rules by virtue of being a small, municipal operation. Cameron was hired in part to help the District comply with the FPC's new planning, data collection, and reporting requirements.

The Forest and Range Practices Act (FRPA) replaced the FPC in 2002. The less-restrictive FRPA provided greater flexibility for licensees, but the scope of information that goes into forest planning and decision making has only increased since then. Better systems are now used by the province to estimate timber volumes, but these are data-rich processes. Satellite imagery and digital mapping increase the accuracy and speed of information but require more specialized skill sets. What the city lacks personnel or equipment to do gets contracted out. Much of the rest falls to Cameron.

• • •

Out in the forest, Brad Laughlin calls ahead on his radio to check if any logging trucks are on their way down. The coast is clear, for now, and Laughlin proceeds up the road in a white pickup with Mission's logo on the door. The road is in good condition, but Laughlin takes his time. As an experienced road builder and foreman of the tree farm, he knows what it takes to damage a road or blow out a tire. Every half kilometre, Laughlin hops on the radio to report his location. If a truck is coming down, he pulls over and waits. The road isn't wide enough to share.

Laughlin knows the harvesting contractors on a first-name basis. He can tell you who is just starting out and who has years of experience under their belt. Sometimes, Laughlin even knows their parents. The harvesters also know Cameron and the others from the Forestry Department, but it is Laughlin they are most likely to hear from out in the woods.

Laughlin has spent the past decade and a half as the foreman on the tree farm. He came to Mission after more than two decades on Vancouver Island doing similar work in the private sector. Like Cameron, Laughlin wears many hats. He locates the timber to be put up for auction, lays out the blocks to be harvested, and decides where the logging roads will go. Laughlin records all of this on a GPS unit and a tablet out in the woods. He sends the data to Cameron, who then makes maps to be sent out to a contractor. The contractor will compile the estimates of timber values and stumpage for the planned cutblocks.

In addition to checking in on the contractors, Laughlin supervises a two-man crew who do about a fifth of the harvesting work. They do the jobs too small to put out for bid. Much of the time, Laughlin's team is out building and repairing roads and, less commonly, deactivating unused ones.

• • •

Over the years, Mission has attracted talented and veteran foresters, technologists, and crewmembers, many of whom, like Pike Cameron (no relation to Kelly) and past foreman Stan Biero, spent decades on the tree farm. For some—like Rocky and Lorne Rockwell, and Howard and Pat Murdoch—it was even a family affair.

The transitions between tree farm managers have happened smoothly in large part because Forestry Department staff have come to the job with such a range and depth of experience. When O'Neal took over in February 2013, Cameron had already been with the department for seventeen years. Laughlin for nine. Fallers George Kocsis and Dennis Klassen had worked on the tree farm for fifteen and twelve years, respectively.

For Chris Gruenwald, who took over as director following O'Neal's retirement in 2019, continuity has been critical to moving forward. "I was very lucky to be coming into a situation like [this], with Brad and Kelly," he said. "Here, it's a lot easier to delegate, to have staff who are super keen and super confident... For me, it's made my job a lot easier."

BCIT practicum student Jillian Wheatley collects coordinates and tree measurements in a replanted area overlooking Stave Lake and Mt. Robie Reid. *Kelly Cameron*

BACARD

The Loggers

TREES ARE TAKEN out of the forest in one of two ways—by ground or by air.

In BC, most timber is cut using ground-based harvesting. This method relies on the use of one-armed, continuous-tracked machines—outfitted with treads for wheels, like tanks—capable of moving methodically across the rough forest floor. Mounted with saws, claws, and hooks, these machines are used to cut, lift, and pile logs. The process today is so heavily automated (and appreciably safer) that operators control most of the action from their ergonomically designed seats in climate-controlled cabs.

At minimum three and usually four pieces of equipment are used. The most powerful machine, the feller-buncher, cuts the trees. Once in position, the machine's mechanical arm, or boom, extends forward, and the vertical, claw-like attachment at the end, called the head, bear-hugs the tree's lower trunk. At the bottom of the head, a whirling blade projects forward and sideways, slicing through the tree. The cab and the boom pivot away from the stand, and the head releases the tree onto a pile nearby.

From there, a second machine, called a delimber, takes over. Using a system of cables and winches that resembles a giant violin bow, the machine will strip the tree of its branches and cut it into desired lengths. Altogether, it takes less than ten seconds to delimb and section a single

FACING Log truck driver Rob Erickson secures a load of logs with cables and chains. *Jason Brawn*

tree. A hoe chucker then moves the logs to a central location, called a landing, where they are stacked into a deck. Yet another machine loads the logs onto a trailer.

While ground-based harvesting is frequently used on the tree farm, aerial harvesting was more common in past years. Aerial logging disturbs less of the forest floor and can be adapted to the steep terrain of the Stave watershed. Two types of aerial systems are used: helicopter logging and tower logging. Helicopter logging, a more recent innovation, is expensive. Tower logging, on the other hand, has been used since the 1950s on the tree farm.

Tower logging employs towers and cables to transport cut trees to a landing along a logging road. The set-up is similar to a gondola, in that taut cables are strung between two points anchored at different elevations. At the high point, the cables are rigged to either a fixed or portable tower. A carriage travels along the overhead cables, moving a dropline and grappling hooks between the tower and harvest area. Once the dropline is in position, the grappling hook is secured to the butt end of the log by a crew member working on the ground. The dropline and log are then mechanically winched uphill to the landing for further processing.

Aerial logging requires more workers to cut, hook, move, delimb, section, and load trees than ground-based systems. The added expense is justified when harvesting high-value timber, such as yellow cedar, and when working in difficult terrain.

• • •

Between the early 1990s and 2018, a local outfit, H&C Logging, harvested most of the timber in the municipal forest. The company held an "evergreen," or Bill 13, contract. In 1991, the province amended the Forest Act to give independent harvesters more security in accessing timber. Prior to that point, operators working on contract for tree farm licensees usually had no guarantee that they would receive new contracts the following year. Evergreen contracts granted access to specified volumes in perpetuity, subject to certain conditions.

H&C Logging had deep roots in Mission. Founders Norm Horn and his brother-in-law, Chuck, were born in Silverdale. Horn spent his life working in area forests. They founded H&C in 1946, and the company worked on contract for nearby mills.

"When the Mission Tree Farm emerged out of nowhere, we were really benefited by our acceptance," Horn said. Their first job on the tree farm was on a block near Rolley Lake in the early 1970s. Afterwards, they regularly bid on Mission's timber. "We would put a bid in and it was always accepted by the tree farm on every block area that was available... We were all successful in... making a profit with what we had agreed to work for." By the time Bill 13 came into effect, H&C had been logging on the tree farm for nearly twenty years.

Small volumes of timber were harvested by the District's own staff or contracted out as special projects. Mission even tried horse logging. In the early 1990s, a local logger, Dale Soper, used a horse-powered skyline system he developed to harvest a small patch near a park and residential area. His team of Clydesdales was entertaining to watch. The method, however, was ill suited for much of the tree farm's challenging terrain.

• • •

After H&C's Bill 13 contract came to an end in 2018, Mission returned to selling trees through timber sales. The city no longer selects bids with a preference for local logging outfits, as there are far fewer of them still in the area. Some contractors hail from the Fraser Valley, but Mission has received bids from as far away as Vancouver Island.

Encouragingly, many of the equipment operators and contractors working on the tree farm are under thirty years of age. They are recent graduates of college forestry programs or came into the business through family connections. After years of shake-ups in the forest industry, Mission's tree farm is again providing opportunities to a new generation of younger workers.

Parks

IN THE FIRST three decades of the tree farm, locals visited the forest in growing numbers, as new roads opened up the backcountry. But they were largely left to their own devices when finding places to camp, fish, or hike.

After 1960, campers who insisted on the modern conveniences of pit toilets and picnic tables could go to Rolley Lake. The lake was pleasant enough, close to town, safe to swim in, and accessible by paved roads. The District once owned most of the land around Rolley Lake. The rest was in a provincial recreation reserve and deliberately left out of TFL 26 when the licence was awarded. In order to create the park, Mission deeded eighty acres of shoreline property to the province in 1960. The BC Forest Service and Parks Branch cleared the land on the eastern edge of the lake the following spring. Workers levelled land for a campground and hauled in sand to create an artificial beach.

Mission and the province collaborated on a second park site adjacent to the tree farm in 1960. Silver Creek Park was located at the base of Bear Mountain, near Steelhead. A sawmill once operated at the base of a small lake, but it had been dismantled years before. The *Fraser Valley Record* reported that the lakeside had become "a popular picnic spot for Coast motorists, tourists, and local residents... Recent Sundays have seen the parking area filled and cars parked along the roadside as well."

Today, locals know the site as Mill Pond. Standing alongside the murky waters of the lake, on a waterlogged lawn speckled with goose dung, it's hard to imagine that the province once considered this a candidate for

Visitors to Rolley Lake Provincial Park can take a short hike onto the tree farm and down to Rolley Falls. *Simon Ratcliffe*

provincial park status. There are no picnic tables, nowhere at all to sit except atop the painted granite boulders placed to prevent cars from being driven or pushed into the pond.

On sunny summer weekend mornings, the parking lot at Mill Pond teems with traffic. Mill Pond is the main staging area for Bear and Red Mountains, the most popular cycling area in the Mission forest. Volunteers from the Fraser Valley Mountain Bikers Association (FVMBA) have named and signed many of the bike trails, and the Forestry Department has planned an interpretive loop trail and boardwalk for around the pond.

At Rolley Lake, most visitors congregate near the beach, setting up barbeques and pup tents. The parking lot, which can accommodate hundreds of cars, is typically full by noon in the summer, as is the nearby campground. A main loop trail hugs the lakeshore, but the more scenic route is a side trail that crosses the park's boundary onto the tree farm and down to Rolley Falls.

Trail Builders

IN THE 1940S and 1950s, who could have anticipated thrill-seeking bicyclists slaloming downhill through the forest, over boulders and across ditches, dodging low-hanging tree limbs, cutting tight corners and then dropping down and over dew-slicked plank bridges to the dirt trail below? And that they would do so on a bicycle outfitted with suspension forks, shocks, hydraulic calipers, disc brakes, and knobby tires mounted on a frame so lightweight that it is easily lofted over one's head?

If you are unfamiliar with "torsional flex," then you are likely not a mountain biker—or at least not one with expensive tastes. Mountain biking, which grew out of a cross between road cycling and BMX riding, fundamentally changed forest recreation. As enthusiasm for mountain biking kicked into high gear, hikers and horseback riders found themselves sharing foot trails with cyclists, whom they often saw as a nuisance or, worse, a danger.

Mountain biking is today a popular, even desirable activity in the forests. But the sport's growth has represented a major challenge for forest managers, as recreational tastes and attitudes evolve. Bear and Red Mountains illustrate what handling these changes looks like on the ground.

• • •

Mission Forestry laid out the tree farm's first planned and properly constructed hiking trail in 1982. Up to this point, hikers either stuck to

FACING Heading up Bear Mountain via the Big Trouble Little Chainring Trail, 2021. *Michelle Rhodes*

logging roads or cut their own trails through the forest. The rugged trail to the top of Mt. Crickmer is a case in point. The Fraser Valley Hiking Club created the route years before the District began signing and maintaining it in the mid-1990s.

Inmates from the nearby Stave Lake Camp dug out the inaugural trail, linking the Florence Lake Forest Service Road to Devils Lake, a popular day use and camping site. The same year, Mission studied the feasibility of creating recreation opportunities at Hoover Lake. Still, the grim financial forecasts of the early and mid 1980s delayed new trail development.

When Kim Allan took over as director, trail construction picked up again. Between 1990 and 1992, new trails went in at Red and Bear Mountains and east from Rolley Lake to Rolley Falls. The Devils Lake Trail was extended and improved. And the Hoover Lake Trail was finally completed.

A few years later, the Stave Dam Forest Interpretation Trail loop became the tree farm's first interpretive trail. On Bear and Red Mountains, trails were upgraded and new ones were built. By the end of the 1990s, a trail along the east side of the Hayward Reservoir, including a side trail down to Cascade Falls, had been completed.

More hikers required more trails, but more trails in turn meant more (often inexperienced) hikers. Mission Forestry prioritized new signage and interpretive displays to educate the public on forest processes and the municipal forest operation. Waste management became a priority, particularly at the lakesides. Vandals repeatedly shot up signs and torched the outhouses to the point where the Forestry Department stopped replacing them.

Mission's forestry staff relied on a growing network of professional and community resources. This included skilled labour provided by the BC Corrections camp at Stave Lake. For the Hayward Reservoir Trail, the District received provincial funding from Forest Renewal BC to hire four displaced timber industry workers. School groups and community organizations volunteered to build trails and help reduce some of the costs of maintenance. For the District centennial anniversary, Mission Secondary students helped resurface Saunders Trail on Bear Mountain.

Starting in 1994, volunteers were also putting in trails for a whole new user group: mountain bikers.

• • •

During the late 1980s and early 1990s, Vancouver's North Shore developed an international reputation among elite riders for its riskier style of free riding. Cyclists hunted for progressively more challenging terrain, with granite outcrops and steeper slopes. Big air was integral to the experience.

Where mountain bike trails did not yet exist, riders built them. They erected homemade ramps cobbled together from scrap wood and dirt. This was before social media, when a thrilling downhill was a secret shared among small groups of avid riders. Mountain bikers in the Fraser Valley put in their own trail networks, including on Bear and Red Mountains.

Bear Mountain tops out at just over 540 metres. This is part of its appeal—taller mountains require more gruelling climbs. The second-growth timber varies widely in age. Riders descend in and out of the forest and recently harvested parcels. The mountain's varied topography allows for several possible combinations of ascents up and runs down. As an added bonus, Bear is the southernmost mountain on the tree farm, close to town, and accessible on three sides by two-lane paved roads.

A handful of cyclists built unsanctioned trails on Red and Bear Mountains in the early 1990s. What started as a rogue activity transitioned into a partnership with Mission's Forestry Department. In 1995, two groups—the Mission Mountain Riders and the Fraser Valley Cycling Club (FVCC)—teamed up with District staff to identify trail routes. One of the riders, Rob Gardner of the FVCC, had extensive trail-building experience, and he put in his own time building challenging routes on both mountains. The improved relationship, though, did not eliminate all unauthorized trails, leaving Forestry crews to pull out those that were dangerous or violated provincial law.

The network of bike trails drew riders from beyond Mission. The FVCC and later the Fraser Valley Mountain Bikers Association organized annual mountain bike races on Bear, starting in 1997 and running most years since.

The numbers of participating riders grew from 250 the first two years in Mission to more than 400 a few years later. A second race event—the Arduum Challenge on Red Mountain—was added in 2007.

The FVMBA's partnership with the Mission Forestry Department was productive but not without its tensions. During the late 1990s, mountain biking in BC was under significant scrutiny because of the legal risks associated with rogue trail building. Conflicts between bikers, hikers, and horseback riders grew. Mountain bikers worried about losing access to parks and forests altogether. In the municipal forest, rogue trails had the potential to interrupt harvesting. Riders, in turn, grew frustrated when logging roads would take out their handiwork.

The mountain biking community sought a place to grow the sport in the Fraser Valley. Mission, meanwhile, had a long way to go in developing Bear's and Red's recreation. Riders put in sweat equity, breaking trail, clearing brush, and repairing bridges. The FVMBA even took on the responsibility of carrying out legally mandated environmental assessments for new trails. In the Mission Municipal Forest, mountain biking trails today are better built and safer than they were twenty years ago.

Hikers and cyclists can and do co-exist, just as biking and forestry do. But it comes with compromises. Trails add more complexity to laying out cutblocks and roads. Logging closures can cut off access. Trail users may spend as much time on logging roads or crossing recently replanted areas as they do under forest canopy. And Mission's trails have been plotted with consideration given after the fact that users would like to look at something other than trees.

Camps

GIVEN ENOUGH TIME, a coastal rainforest has a remarkable ability to heal itself after shock. An ice storm or blowdown, a clearcut or forest fire. Sunlight pours through the open canopy, down to the forest floor. Fallen trees provide habitat for species well adapted to disturbance. New seedlings take root. The process of renewal begins.

Forests also nurture people. Away from the noise and light pollution of the city, forests offer respite from a digitally connected world. Engaging with the forest can be a deeply cultural and social act. For some, the mere fact of being away from home and routine can be its own reward.

Over the past few decades, non-profit organizations in BC and Canada have worked with communities, local and provincial governments, and donors to create dedicated facilities and experiences for underserved populations. Three organizations in particular have established facilities or expressed interest in doing so in the forests west of Stave Lake: the Zajac Foundation, the Tim Horton Children's Foundation, and the West Coast Kids Cancer Foundation (WCKCF).

The interest in Mission stems from its geography and community relationships. The municipality is only a ninety-minute drive from Vancouver, and the forest is close to hospitals and other services. Most importantly, organizations have also found enthusiastic partners in the District, its Forestry Department, and neighbouring First Nations.

Celebrating the grand opening of Zajac Ranch's Aboriginal Arts and Education Centre in 2017. *Terry Hood*

The Zajac Ranch

The Mel Jr. and Marty Zajac Foundation operates the Zajac Ranch for Children, the longest-running all-abilities camp facility in the Fraser Valley. Mel Zajac, a property developer, building contractor, and long-time philanthropist, along with his wife Irene, established the organization in memory of their two late sons.

The forty-acre Zajac Ranch is located on the site of the former corrections camp at Stave Lake. The foundation purchased the land and buildings of the camp a year after it closed in 2002. It updated the dorms, cafeteria, and programs building to accommodate the specialized needs of camp attendees. The foundation built a health centre, swimming pool, rentable meeting space, and equestrian and children's zoo facilities, and it continues to incorporate new structures as funding and demand warrants. Recent additions include a ropes course, treehouse camping, and an outdoor amphitheatre.

Since opening in 2004, the Zajac Ranch has hosted thousands of children and their families for single- and multi-day camp excursions. The facility is open year-round but sees most campers during a ten-week summer season. In the off-season, the Zajac Ranch supports K–12 classes and children's groups of all abilities. It generates additional revenue through facility rental and overnight accommodations.

The Zajac Ranch is surrounded on three sides by the tree farm, but it is privately held and not part of TFL 26. The Zajac Foundation and Mission's Forestry Department work together on shared needs, such as maintaining the FLFSR for bus and small vehicle travel. The Zajac Foundation has also actively participated in Mission's long-term planning processes for the west side of Stave Lake.

Tim Horton Children's Foundation

In 2010, the Tim Horton Children's Foundation first proposed the idea of building one of its destination summer camps near Mission. If built, the project would be the foundation's first in BC and would primarily serve children from low-income households. The foundation tabled its plans a few years ago, but it has maintained its Licence of Occupation on the proposed site.

Like the Zajac Ranch, the Tim Horton's project would be located just outside of Mission's tree farm at Pine Lake, on the grounds of the former Boulder Bay Camp. The proposed year-round facility would accommodate up to three thousand youth annually. The foundation laid critical groundwork, including archaeological and land surveys, budgeting, and site planning, and obtained a licence. The District of Mission, meanwhile, commissioned a feasibility study for the project.

The most immediate challenge, however, has been access via the FLFSR. North of the Zajac Ranch, road conditions are less suitable for bus travel. Beyond the turnoff for the Sayres Lake Campground, high-clearance vehicles are required, and year-round access for buses and passenger vehicles would require significant road improvements around Sayres Lake. As of 2018, an additional $3.5 million was needed to extend utilities to the camp.

West Coast Kids Cancer Foundation

In 2017, the West Coast Kids Cancer Foundation approached Mission's director of forestry Bob O'Neal with a request and a proposal. Would the District consider letting the foundation locate a permanent oncology camp for kids and their families within its community forest?

The WCKCF presented a vision of a year-round facility built to maximize the recreational potential and therapeutic power of the forest. Mission and neighbouring First Nations have encouraged the development of projects along Stave Lake like this, projects that are family-oriented, respectful of Stó:lō values, and complementary to other land users.

The proposed site for the facility was initially Rockwell delta, just south of Devils Lake. The gently sloping parcel of land overlooks a small, scenic bay along the Stave Reservoir. The Kwantlen and the WCKCF submitted a joint application to remove the land from TFL 26. In preparation, an archaeological survey was completed and a new bridge constructed in 2018. While the joint application to the province was still awaiting review, WCKCF has been in discussions with the Forestry Department about possible alternative locations.

Bridge Builders

FOR MUCH OF the tree farm's history, the District and the province functioned as though they were the only ones with legal claims to the lands around Stave Lake. A series of rulings by the Supreme Court of Canada changed this.

In *Calder v. British Columbia* in 1973, the court recognized the existence of Aboriginal Title, or the right of Indigenous Peoples to live in and use their ancestral territories. The *Calder* case opened the doors to additional legal challenges by Indigenous Nations. In the years that followed, the court reaffirmed, defined, and strengthened the legal standards for these rights. These rulings applied in particular to those Nations that had never signed a treaty with the colonial or federal government. All of the Fraser Valley and most of BC is unceded, or lacking treaties.

For land-use managers, including those operating tree farm licences, *Calder* did little to change day-to-day operations. The 1997 *Delgamuukw* decision, in contrast, reshaped government-Indigenous relations in forestry. The court determined that the Indigenous right to title extended to determining how unceded lands in traditional territories get used. A First Nation is not limited to hunting, fishing, and other historical practices. Just like private landowners have great leeway in how they use their property, so do Indigenous Peoples. They may, for instance, choose to participate in commercial resource development.

The *Calder* case opened the doors to additional legal challenges by Indigenous Nations. In the years that followed, the court reaffirmed, defined, and strengthened the legal standards for these rights.

The court's ruling meant that governments and resource users were required to consult local First Nations on land-use decisions in order to operate on unceded Crown lands. Tree farm licensees had to seek input from their Indigenous neighbours on forest management plans, cutting permits, and licence renewals. After *Delgamuukw*, the Kwantlen First Nation stepped forward to exercise their right to be consulted on Mission tree farm plans. In response, Mission began corresponding with the Kwantlen and the Stó:lō First Nations for review of proposed management and forest development plans. By the mid-2000s, the District was soliciting feedback from several more Nations, including the Matsqui, Sumas, Katzie, Skawahlook, and Seabird Island Indian Band.

Still, the courts had left open the question of what constituted meaningful consultation and did not equate it with consent. Resource managers typically sought consultation after plans had already been drafted and preliminary decisions had been made. Accommodations would be made for archaeologically significant sites or in response to other concerns, often after the fact. Consent to proposed activity was not required. Mission was little different from other licensees during this time. To meet these obligations, forest manager Bob O'Neal and forest technologist Kelly Cameron met with Kwantlen representatives Tumia Knott and Tony Dandurand once or a twice a year to review proposed cutting permits. The relationship and the meetings were, according to Knott, "very clinical."

Bigger questions—how, for instance, to move towards common values and principles, or what it means to share in the benefits from the TFL—were left unanswered. The Kwantlen and the District agreed to address some of the heritage inventory needs, including by bringing on Kwantlen field workers to complete archaeological and cultural site assessments. The Kwantlen had also worked with BC Hydro on critical cultural inventory research.

The Kwantlen pushed for a different process, one in which they and other First Nations would be a part of planning and decision-making processes at a much earlier stage. Getting to this point, to a relationship that is based on working collaboratively and with mutual respect, would take time. The District, meanwhile, was still required to consult on cutting permits, management plans, and other documents on a regular basis.

The shoreline of Devils Lake. *Terry Hood*

Matters came to a head over the replacement of Tree Farm Licence 26 in 1999. The Kwantlen, disappointed with the failure to move beyond the limited type and scope of consultation, met frequently with Mission's forestry staff and the Ministry of Forests to argue for a different approach. Standing before Mission's mayor and council, Chief Marilyn Gabriel articulated her frustration that a licensee operating in her Nation's traditional territory had excluded them from meaningful involvement in key decisions. Among these was the decision to pursue another twenty-five-year licence.

The Kwantlen wanted the next licence to include language that recognized a stronger role for the Indigenous partners and communities. Knott explained that the Nation's main message was that "the future we would like to see is one where we are here on the landscape with you." The Kwantlen made it known that, absent changes in the language and an authentic commitment towards an inclusive management process, the Nation would not support the licence's renewal.

The Ministry of Forests approved the licence renewal without the asked-for language. The Kwantlen filed a legal challenge to the renewal, requesting a judicial review of the process resulting in the renewal of Mission's tree farm licence.

The relationship between the Kwantlen and the District sank to a new low. Mission's political leadership had little experience working with First Nations at a government-to-government level. Council members and the mayor's office were openly skeptical of the Kwantlen's requests. Neither party fully understood the other's motivations. "We weren't trusting of what their intentions were," recalled then forestry manager Bob O'Neal, "and they weren't trusting of what our intentions were."

During the licence renewal period, the Kwantlen had also reminded the province that they, the Matsqui, Leq'á:mel, Katzie, and other Nations had been denied revenue opportunities in the territory. And so the Kwantlen asked for half of the revenue from TFL 26. While neither this nor the requested changes to the TFL came to be, BC's Ministry of Forests did announce that it would issue a First Nations woodland licence to the Kwantlen First Nation. The woodland licence helped ease some of the tensions between the Kwantlen, the District, and the ministry, while the Kwantlen staffed up to run their own woodlot operation on Blue Mountain.

Kwantlen's Elders reiterated that the Stave area is the spiritual heart of the Nation. The District would need to engage Indigenous communities with a good heart. Knott, Dandurand, and Drew Atkins from the Kwantlen First Nation met frequently with O'Neal and others from the District. Mission Forestry offered assistance to the Kwantlen as they readied plans and materials for their woodlot. Slowly, if only on a person-to-person basis at first, bridges were being built.

In 2008, representatives from Mission and the Kwantlen gathered at Zajac Ranch. Mission Forestry arranged a tour of the tree farm for Kwantlen Elders and councillors. At Sayres Lake and at the Foreshore Flats, the Elders found spent shotgun shells and tire tracks across the exposed lake bottom of the Stave Reservoir. Culturally significant sites were visibly at risk. In the years that followed, the District, the Kwantlen, as well as the Leq'á:mel, Matsqui, and other stakeholders, would find common purpose in transforming the west side of Stave Lake.

9 THE WILD STAVE WEST

ANYONE WHO SPENT time on the west side of Stave Lake in 1990s or early 2000s likely has stories to tell. They might have witnessed (or participated in) target shooting, bonfire burning, driving through creeks, garbage dumping, or worse—vandalism, drunk driving, torching stolen vehicles, theft. Many locals avoided the area altogether.

Getting a better handle on crime and unregulated activities was the first step in reimagining the area as Stave West, a multiple use, family-friendly destination. Change was needed to ensure the safety of Mission's logging operations, but it was even more important to Indigenous Peoples

PRECEDING Stave West Leadership Team members and stakeholders paddle the Stave. *Jason Brawn*

FACING Remnants of a weekend bonfire on the Foreshore Flats. *Jason Brawn*

whose ancestors shared this territory. Illegal and unsanctioned uses were disturbing archaeologically and culturally significant sites.

Achieving a shared goal of transforming this area would require a new framework for consultation and engagement between Indigenous Peoples and the District. Other stakeholders in the area—BC Hydro, recreation groups, schools—would need to be at the table. Together, they would help shape a collective vision of what is possible, desirable, and sustainable in Stave West.

Security cameras show this pickup on the FLFSR with an old couch in the back. Dumping and waste-burning are major problems, with high costs for the municipality. *City of Mission*

Patrolling the Stave

IN THE EARLY 2000s, the Mission Tree Farm had a gun and stolen car problem. Over the previous decades, vehicle theft surged in BC, especially in the Fraser Valley. A disproportionate number of these cars were being dumped and torched along the Florence Lake Forest Service Road. In just one week of April 2006, RCMP officers on patrol located eleven abandoned vehicles over a twenty-kilometre stretch of the FLFSR.

The financial and environmental costs of this criminal activity were climbing. Vehicle fires could easily spread to the surrounding forest. Marijuana growers dumped used soil, plastic barrels, and other waste materials off the side of logging roads and parking areas. Pickups loaded with household waste did much the same. Shooters nailed targets to valuable firs and cedars and left the trees riddled with bullet holes. The District spent tens of thousands of dollars each year replacing equipment and signage, fixing gates, towing stolen vehicles, cleaning up waste, and writing off stolen cedar blocks and damaged trees.

Mission's proximity to rapidly growing urban populations was part of the problem. More people were using the forests to the west of Mission. This, in turn, pushed the more unsightly or undesirable activities eastward into the comparably remote lands west of Stave Lake. The FLFSR was also ungated. Via the service road, the public could access thousands of hectares of forests and lakeshore.

There were simply too few eyes on too much territory. Most lawful visitors congregated at a handful of spots—Devils Lake, Sayres Lake, and the BC Hydro boat launch. Anyone determined to cause trouble would drive further up the FLFSR and away from popular spots. The problem worsened after the sun went down. Regularly gating the road would have created logistical problems for the agencies and contractors with legitimate business on Crown land. The District and the provincial government would have faced huge public backlash if they were to restrict access.

What Mission did do is monitor who was coming in and out of the forest on the FLFSR. In 1995, the District of Mission, BC Hydro, and BC Corrections (which at the time still operated a camp north of Devils Lake) teamed up to cover the cost for two hundred hours of RCMP time. For several summers, RCMP officers set up irregular roadblocks along the road; they soon added occasional bush patrols. Mission's Forestry Department also hired a security firm to conduct weekend checks at the bottom of the FLFSR.

Mission further expanded the monitoring program in 1999. It built a gatehouse at the entrance to the FLFSR and installed a gate for use during periods of high fire danger. It contracted a security firm for evenings and weekends, seven days a week, between mid-May and mid-November. The provincial vehicle insurance agency, Insurance Corporation of British Columbia, provided grant funding to cover much of the increased cost.

Although ICBC's funding of the program was predicated on reducing stolen vehicles, the effects were wider ranging and detectable immediately. In the four years the program ran, reports of vandalism and theft dropped, and visitor use increased. When ICBC funds dried up, Mission could only afford to pay for some of the monitoring. Speaking to the *Mission City Record* in 2003, forestry director Kim Allan expressed frustration over other users not sharing the cost. "Only the District of Mission seems to want to do something," he said. "We're trying to focus on other partners, but they only seem to be interested in their own site, not the bottom (of the road)."

Relief arrived in 2006, when the District council agreed to cover the costs of a new RCMP position, a "beat cop" whose territory included the FLFSR. The change in funding also recognized that security along

Efforts to combat illegal activities over the past fifteen years have included more police patrols and increased signage. *Terry Hood*

this road was a "Mission problem," not just a "Mission Forest problem." Anecdotally at least, according to the Forestry Department, the RCMP's increased presence had a deterrence effect.

Still, curbing criminal behaviour west of Stave Lake required more than the added RCMP patrols and contracted security crews could provide. The police could not be everywhere at all times. To boot, poor cell service made it nearly impossible for visitors to report problems as they occurred.

A more holistic approach was needed. To that end, Mission began imagining Stave West.

Planning the Stave

BY 2008, THE District had already spent more than a decade trying to tamp down illegal activity in the municipal forest. Building a more visible security presence along the Florence Lake Forest Service Road helped, but it was difficult and expensive to sustain.

Each summer was busier than the last, but many families and other recreational users—worried about having their cars broken into, shot at, or worse—stayed away. The area lacked planned, maintained, and regulated campsites, trails, and bathroom facilities. The potential for conflicts between different groups escalated, especially between motorized and non-motorized recreation. Uncontrolled camping, campfires, and off-road vehicle use compounded damage to culturally sensitive sites and the natural environment, issues that cost the District and the province time and money to remedy.

A more substantive solution required a change in perspective. Instead of asking whether increased security could support increased recreation, Mission began asking whether increased recreation could make the forest safer.

• • •

When the District first decided to take a different approach to management on the western half of the TFL, it did what government agencies usually do. It hired consultants.

Mission's council and Forestry Department recognized the untapped potential for tourism west of Stave Lake. They envisioned being able to offer visitors experiences ranging from canoe rentals to zip trekking. But the area's poor reputation and history of undesirable activities made it financially risky to pursue this type of development. And any new facilities, trails, or recreation sites would have to be built while taking into consideration the location of future timber harvesting.

A consultancy firm carried out a recreation and tourism feasibility study, the *TFL 26 Recreational Opportunities Feasibility Analysis*, in 2009. It was followed by an employment study the following year. Then, in 2011, the province granted "interpretive forest" status to the western half of the tree farm. This, the District hoped, would open up new opportunities for partnerships and branding.

On the heels of the two reports, Mission's forestry team recognized that a reactive, top-down process would produce short-lived gains, like cleaning up a shoreline or reducing shooting in one area. These gains, however, would be expensive to maintain without broader public support and involvement. In order to work, comprehensive planning for the area would have to be collaborative. Creating lasting and sustainable change in the community forest required building trust and repairing relationships with Indigenous partners.

To Mission, the escalating misuse of the area over the previous two decades had become a long-term problem. To the Kwantlen, Matsqui, Leq'á:mel, Katzie, and other Nations, crime and misuse was only the latest stage in a broader process of dispossession and disconnection from their traditional, shared territories.

The District's relationships with local Indigenous Nations ranged from nearly non-existent to, at best, progressing but strained. The Kwantlen had pressured Mission to engage in more meaningful consultation and involvement in the forest's management. The Leq'á:mel would soon do the same in regard to new development in Hatzic Valley, on the east side of Mission.

As momentum built around the vision of revitalizing the west side of Stave, the Mission Forestry Department met frequently with Tumia Knott

Lasting and sustainable change in the community forest required building **trust and repairing relationships with Indigenous partners.**

and Drew Atkins of the səyem̓ qwantlen business group. Kwantlen Elders visited Stave Lake, touring municipal operations and sharing with staff the meaning and significance of this place. At the outset of the planning process, open and honest communication was essential, according to then forestry manager Bob O'Neal. Above all, the Kwantlen emphasized the importance of being patient, deliberate, and inclusive. "If we... were pushing along a timeline, we'd hear about it," explained O'Neal.

• • •

To help with the Stave project, the Forestry Department hired Terry Hood, a community tourism planner. Hood was finishing a project for the Mission Regional Chamber of Commerce, assessing the District's tourism potential. Serendipitously, Hood and O'Neal ran into one another at a rural tourism conference in Kamloops. O'Neal was there to talk about the tourism and recreation possibilities of the Stave area.

O'Neal arranged for Hood to come on as an adviser to the Stave West project. "Terry was phenomenal in that he knew exactly where things could go," recalled BC MLA Pam Alexis, who was a member of the chamber at the time. "Bob had the vision and the drive and the passion."

The circle of people involved in the project grew from there. "People realized that in order for this project to be successful, it had to be more than the District and the Nations," noted Hood. "It really had to involve the many other key groups that are contributing today." The RCMP were a part of the initial conversations. They were soon joined by representatives from the Mission Regional Chamber of Commerce, BC Hydro, and Mission public schools.

The Forestry Department arranged site visits to build support for the planning process. School administrators, local post-secondary institutions, and recreation organizations were invited on tours of the Stave West area. In planning sessions held at Zajac Ranch, groups were asked to contribute ideas on how they would use the area or what they would like to see happen. According to Hood, "We were just introducing people to this beautiful area and the possibilities."

By 2014–15, the Stave West planning team counted more than thirty members. Jason Thompson and E. Alice Thompson of Leq'á:mel attended the first meeting, and Cindy Collins of the Matsqui First Nation joined in soon after. The Ministry of Forests, Lands, Natural Resource Operations, and Rural Development sent representatives to participate, as did the University of the Fraser Valley, the BC Institute of Technology, and the Aboriginal Tourism Association of BC.

The process relied heavily on relationship building and frequent communication but not on any prescribed protocol for community consultation. No firm deadlines were set, at least not in the beginning. Alexis reflected on the sheer number of people involved: "Because there were so many players, I was worried about getting everyone on the same page. I was worried about the buy-in from everybody."

At the time, O'Neal shared a similar concern, wondering if the province would recognize the value of the process. "I don't think they really understood where we were going," O'Neal recalled. "It was almost pre 'it's time.'"

• • •

In 2015, the planning team released the *Stave West Master Plan*. The report identified ten guiding principles that would be used to determine future needs and projects. Top among these was the need to honour the meaningful relationships that had been established. The other principles included effective communications; protecting culture and heritage; investment and economic benefits; education and interpretation; stewardship; respectful partnerships; safety; family-friendly recreation and tourism; and economic diversification and enhancement of operations in the licence area.

Even before the plan's release, the partnership had generated concrete results. The provincial and federal governments funded road upgrades to Zajac Ranch. These improvements made it easier for school buses to travel to and from the facility. In 2013, Recreation Sites and Trails BC, a provincial agency, completed work on the first organized campground in the municipal forest—a forty-site campground at Sayres Lake. The sәyeḿ qwantlen group held the contract to manage the campground during the

At the Stave West Leadership Team monthly meeting, this time hosted by the Matsqui First Nation. *Terry Hood*

summer. The District met with K–12 educators and local universities to transform Stave West into an outdoor learning laboratory.

The plan lacks hard targets and deadlines, but it is nonetheless ambitious. According to Hood, the plan was "the first tangible evidence that different groups were working together towards a common vision."

It had taken Mission Forestry and the planning team seven years to get to this point. Grants and internal funding covered room rentals, consultants, and coffee but not the patience and mental energy required to keep the whole process afloat. Relationship building grew from the gentle but persistent nudging by O'Neal and Hood and reminders from Knott and Atkins to take the time needed to do the work right and the right work. Most of all, the parties at the table had to be willing to listen intently to one another, build trust, and to focus on what is possible.

Stave West is being transformed into a space safe that is welcoming for families. *Jason Brawn*

Down to Details

THE FLORENCE LAKE Forest Service Road is busier than ever, even mid-week. Hikers and their dogs can be seen travelling by foot along the road between Rolley Falls and Devils Lake. The campsites at Sayres Lake, Kearsley Creek, and Rocky Point are full and many are booked well in advance. A regular crowd of quad and dirt bike riders congregate near the turnoff to Foreshore Flats. Young riders take the lead, their parents tailing close behind.

And yet the forest is quieter. The target shooters have largely left the area. More of the drivers on the FLFSR are novices on logging roads and they proceed with greater caution. The weekend rowdies—the ones who cause the most damage, who burn pallets on the waterfront and leave trash piles behind—are still there. The pickups fully loaded with old furniture and garbage are too. But they show up later in the evening, hoping to avoid detection from the RCMP or contracted security teams. Greater vigilance has worked; there have been fewer RCMP callouts to the area since 2016.

These positive changes stem from the Stave West planning process and the tangible steps taken as a result. Regular RCMP patrols and weekend security, road upgrades, and the construction of four campgrounds and more than two hundred campsites have made the area safer and more accessible. As new residents pour into the Fraser Valley and eastern sections of Metro Vancouver, the demand for what the Stave West Forest and Recreation Area has to offer has never been higher.

• • •

Once a month, the members of the Stave West Leadership Team come together to discuss ongoing and proposed project needs for the SWFRA. The *Stave West Master Plan* called for the creation of a tripartite board—including representatives from provincial, municipal, and First Nations governments—to manage the area. The leadership team fills this role for now.

Team members participate in information sharing and collective decision making. The team is non-hierarchical—there's no "boss," per se, although each team member oversees or reports back on projects specific to their organization or Nation. The RCMP reports on patrols along the FLFSR. A BC conservation officer might provide news on bear activity in the area. Tom Blackbird, the provincial district recreation officer whose jurisdiction includes the Crown lands of the SWFRA, shares updates on the status of requests or applications for new recreational facilities or accommodations. Blackbird was also instrumental in getting the campgrounds planned and constructed so quickly.

At the table are many of the same people who were involved in developing the master plan: Cindy Collins (Matsqui), Tumia Knott and Drew Atkins (Kwantlen), Kelly Cameron (Mission Forestry Department), Terry Hood, and Director of Forestry Chris Gruenwald, who came on board when Bob O'Neal retired in 2019. Gruenwald previously worked with the Kwantlen and Katzie forest tenures, so he came to the team already familiar with the process. Representatives from the Leq'á:mel have rejoined the group after some time away. Until recently, the Stave West planning process was guided by long-time team member and Mission's manager for community engagement and corporate initiatives Michael Boronowski; Clare Seeley, the city's manager of tourism, took on this role in 2020.

The continuity in membership has helped the team to carry out the master plan's recommendations and to coordinate input from a range of recreation-based clubs using the SWFRA. By 2021, the Stave West Stakeholders Group had grown to include more than four dozen organizations, schools, and agencies. Prior to the COVID-19 pandemic, the leadership

As new residents pour into the Fraser Valley and eastern sections of Metro Vancouver, the demand for what the Stave West Forest and Recreation Area has to offer has never been higher.

team invited all members of this group to a general meeting twice a year. These meetings were an opportunity for stakeholders to network with each other, express their ideas and provide feedback on proposed changes, and hear what other groups are doing. Individual organizations may also work one-to-one with Stave West partners on special projects.

While on the surface these processes may seem cumbersome, they are entirely consistent with the vision for the SWFRA as a safe, inclusive, family-friendly, and sustainable area. Stakeholders inform decisions on locating and developing new trails and facilities. They provide first-hand knowledge of what their members want, such as the features of a good horse trail. Some provide labour for clearing trails, building corrals, organizing clean-up events, and repairing damage done by other users. Simply put, the more that community members and recreation groups feel heard, the more likely they are to provide input or work to improve conditions on the ground.

Just as important, a more inclusive, collaborative consultation process helps to reduce user conflicts and damage to environmental and cultural assets. This is especially important in the Stave watershed, given its rich multicultural heritage and its importance to the Kwantlen, Leq'á:mel, Matsqui, and Katzie Nations.

In 2018–19, the leadership team invited stakeholders to submit proposals that outlined what they would like to see develop in the SWFRA. Some expressed an interest in specific locations in the forest, such as high-country logging roads (snowmobiling), boat launches (kayaking), or trails (hiking, horseback riding). Others proposed far more ambitious training and education programs (off-road driver training) and site developments (facilities and adaptive camps). The requests were then compiled by Cameron, Mission's forest technologist, into a single map, to help the leadership team visualize recreation corridors and possible problem areas.

Already, this integrative recreation planning process has resulted in the construction of two new trails, a fire lookout, and an interpretive signage plan for Stave West. Situated on a ridge just west of Devils Lake, with panoramic views of the Stave Valley and Mt. Baker (elev. 3,286 metres), the lookout can today be reached by a new six-kilometre trail from Seventynine Creek. This trail is only the second day hike built by the District on the west side of Stave Lake, the other being the more arduous full-day climb up Mt. Crickmer. Another trail, constructed in 2021, encircles Devils Lake. To help guide visitors and educate them on the history and management of this region, the first group of new kiosks and interpretive signage (designed with help from students at the University of the Fraser Valley) have been installed along the FLFSR, with more planned.

• • •

The most recent rounds of stakeholder consultations also shone light on potential user conflicts, as was expected. Just as important, the process also illustrated where some shared interests lay. Many spoke of the opportunities for education, interpretation, and economic development. All wanted safe spaces to play. Stakeholder-generated ideas also aligned well with the principles of the planning process, with one glaring exception: the Foreshore Flats.

The Foreshore Flats

THE MODERN SHORELINE of Stave Lake is much longer because of the Stave Falls Dam. When it was built a century ago, the dam raised lake levels. When it was raised again only a few years later, the lake expanded outwards. These waters inundated culturally and ecologically significant sites. Some of the ancestral Stó:lō history is entombed in the mud and water.

The westernmost portion is more reservoir than lake. BC Hydro raises and lowers the level of the reservoir according to seasonal precipitation forecasts and energy demands. The utility impounds water in the summer, releasing it slowly as temperatures rise and Vancouver turns on its air conditioners. The lake is at its lowest level in autumn. Lake levels start to rise again when late November rains and winter snows return. By late spring, more water is released from the dam, and the lake falls once again. As it does, the lake shrinks, its waters pulled farther from the shore.

When this happens, an expansive stretch of reservoir bottom is exposed at a site called the Foreshore Flats, located between Kearsley and Sayres Creeks. Little vegetation grows on the Flats. Rainwater collects in shallow pools atop the bare sands.

Jacked-up trucks, quads and Jeeps, diesel bikes, unloved cars, and old vans descend on the Flats each fall and spring. The Flats make for good mudbogging—driving hard and fast across packed dirt and through waterlogged sands. Some four-by-fourers deliberately try to get stuck, simply for the fun of trying to get unstuck. Fellow drivers come to the rescue to push, tow, or winch the fallen out of the muck.

Most visitors just come to watch, mesmerized by the theatre in the dirt. The Flats draw hundreds, even thousands, of mudboggers on a day when conditions are optimal. Many riders find a sense of community and adventure here. Some loosely claim membership in the "Dirt Church," while others come to party late into the night around bonfires of pallets, cardboard boxes, discarded mattresses, and scavenged wood.

Everything about what is happening on the Foreshore Flats is contentious. The activity has until recently been entirely unsanctioned by any authority and at best tolerated by the province and BC Hydro. Mudbogging and bonfires damage shoreline ecologies and deter cultural uses of the site. On dry days, vehicles kick up so much dirt that dust clouds can be seen from as far away as the Stave Falls Dam, eleven kilometres to the south. And then there are the myriad of safety and liability issues.

A not-insignificant minority of visitors to the Flats create a disproportionate number of issues—leaving trash behind, burning toxic materials, riding through patches of vegetation and into surrounding areas. Parties on the sands attract teenagers who come to hang out and drink and then drive at high speeds on the Florence Lake Forest Service Road.

The Four Wheel Drive Association of BC has advocated passionately and persuasively to keep the Flats open to mudbogging. The association is active in remedying some of the worst effects of uncontrolled use, such as cleaning up garbage problems at the site. Because they are most present in the off-season, members of the association and more recently the Dirt Church are often the "eyes and ears" on the ground, reporting damage done, fixing gates, and speaking to other off-roaders when they're riding where they shouldn't.

If not for advocates like this, the Flats may well have been closed to motorized recreation years ago. A partial closure is in place but in exchange for a designated area for mudbogging. Whether that is enough to sate the appetites of thrill seekers is yet to be seen. Mission Forestry and provincial agencies continue engaging with many different stakeholders—some of whom are unfazed by the mudboggers, while others think it is well past time they leave. Where they would go is an equally vexing question. Many are simply going to the east side of Stave Lake, beyond the boundaries of the Mission Tree Farm.

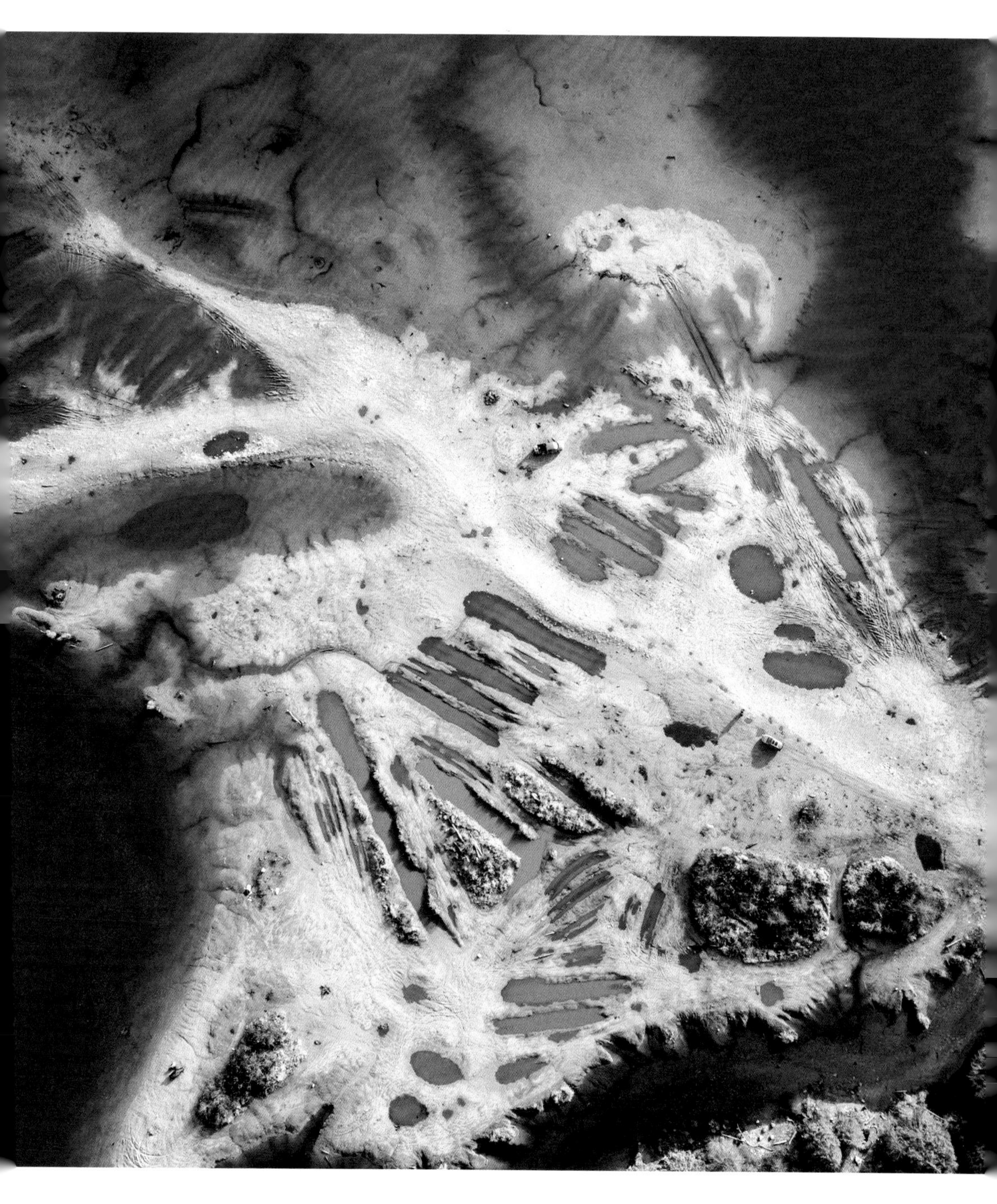

The mud flats seen from overhead, with deep grooves etched by mudboggers filled with lake and rainwater. *Jason Brawn*

The Blanket

FOLLOWING ITS DECISION to replace Mission's Tree Farm Licence 26 over the objections of the Kwantlen, BC's Ministry of Forests pledged to expand Indigenous forest management in the area. This included allocating a First Nations woodland licence to the Kwantlen First Nation on Blue Mountain, west of Stave.

An FNWL resembles a tree farm licence, in that it is renewable and requires that licensees to develop management plans and carry out a certain level of harvesting. After signing an interim agreement with the province in 2006, the Kwantlen received their licence for the woodlot on Blue Mountain, just west of Mission, in 2011. The licence area is small—only eight hundred hectares—but the woodlot gave the Kwantlen an independent toehold in forestry. The Kwantlen established a separate business group, səyem̓ qwantlen, who then worked with a contractor to develop harvesting and long-range management plans. The Nation has created strategic forest reserves on the woodlot to protect ecological and cultural values. The woodlot has also provided career opportunities for Kwantlen youth.

Prior to the woodlot's development, the Kwantlen's neighbours, the Katzie First Nation, had planned to apply for a community forest tenure along the eastern shores of Alouette Lake. The Katzie and Kwantlen instead decided to jointly apply for an FNWL. A partnership was considered advantageous to both procuring and managing a tenure. In 2019, the two Nations received the Katzie-Kwantlen FNWL, known as the K&K. For the Katzie, the journey to local control over forests in their territory has taken nearly fourteen years.

The K&K includes 6,732 hectares of forest that run between the shores of Alouette Lake and the western boundary of the Mission Tree Farm. The tenure is more than eight times larger than the Kwantlen woodlot and 30 percent larger than the Stave West Forest and Recreation Area. In creating the licence area, Mission and the Ministry of Forests agreed to a land swap. Totalling close to 360 hectares, two parcels of productive forest were removed from TFL 26 and added to the woodland licence. The parcels were already well connected by old roads to the K&K. In exchange, the province added several timber berths to TFL 26. These lands were surrounded by or immediately adjacent to Mission's forest tenure. The land trades greatly improved efficiencies for harvesting and ecological planning for both licensees.

The Nations have begun looking at how they can train and equip Kwantlen and Katzie youth to work in the forest. One strategy is through the creation of an Indigenous Forest Guardianship program to serve on the K&K, at the Kwantlen woodlot, and in the SWFRA. The Leq'á:mel have started their own Guardianship program, with its members working on the lands east of Mission. Indigenous forest guardians are knowledge keepers and gatherers, helping to protect cultural and ecological resources and educating the public on sustainable and culturally sensitive practices. The knowledge they accrue from working in the woods can also help to inform community-led planning and forestry practices.

The Matsqui and the Leq'á:mel have also pursued forestry opportunities, but they have not as yet secured a long-term licence. Along the north shore and east of Stave Lake, most forest land is managed not through area-based tenures, like tree farm licences, but by BC Timber Sales (BCTS). The BCTS is a Crown corporation charged with planning harvesting activities and sales of timber through a competitive process. Most of the forest surrounding the Leq'á:mel Nation's reserve lands is administered by BCTS.

Both the Leq'á:mel and the Matsqui previously signed agreements with the province that would grant them access to non-replaceable licences in

Indigenous forest guardians are knowledge keepers and gatherers, helping to protect cultural and ecological resources and educating the public on sustainable and culturally sensitive practices.

areas managed by BCTS. Unlike Mission's licence, which can be replaced after twenty-five years, a non-replaceable licence covers a specified volume of harvesting over a five-year period. The focus is on generating revenues through timber extraction, not long-term management in accordance with cultural practices and values. Approximately 21 percent of what was harvested in 2018 in the South Coast Natural Resource Region (including the Fraser Valley) was harvested under non-replaceable licences allocated to First Nations. One of the licences that year was held by the Matsqui.

In 2017, the Leq'á:mel signed on to a Forest Consultation and Revenue Sharing Agreement with the province. The agreement spells out when and to what level the Nation is to be consulted on forest plans and licences, including TFL 26, within its traditional territory. The Leq'á:mel, Sumas, and Skawahlook First Nations are also party to a treaty negotiation process, now in its final stages. The eventual outcome of these negotiations may further reshape Indigenous forest tenure north of the Fraser River.

• • •

The Kwantlen took an early role in developing a working relationship with Mission's Forestry Department. Because of this, Mission and the

Nation now collaborate frequently on forest-planning initiatives, including the SWFRA. Being first, however, should not be misinterpreted as somehow winning the race to stake a claim to territory.

The Stave watershed is part of the shared territory of many Nations, none of whom have ceded rights to the space in a treaty. The Leq'á:mel have given notice to the province and to Mission that they intend to assert their rights, title, and responsibilities to territory that extends to the north shores of Stave and Alouette Lakes and west to Silver Creek in Maple Ridge. To that end, both the Leq'á:mel and the Matsqui First Nations are at the SWFRA planning table, alongside the Kwantlen. Collectively, and as individual Nations, they have pushed for a space for dialogue and meaningful consultation.

Building these relationships with First Nations partners was a priority for the Mission Forestry Department under director Bob O'Neal. His successor, Chris Gruenwald, continues to do the same and is in a unique position to do so. Before coming on as Mission's new director of forestry, Gruenwald and his business partner consulted on the Kwantlen woodlot, the K&K, and the Cascade Lower Canyon Community Forest in Hope.

• • •

A blanket of forest stretches across more than 250 square kilometres of mountains, lakes, and creeks, from Alouette Lake, across Blue Mountain and Mt. Crickmer, and down across the southeastern shore of Stave Lake. Like a quilt constructed of irregularly sized cuts of fabric, the adjoining tenures on the Fraser Valley's north shore have been stitched together to create this blanket. First Nations manage two parcels, local government one, and two small pieces are held by forestry training programs at BCIT and UBC.

Over time, each piece in this patchwork will become more distinct, shaped by culture, land use, boundaries, and topography. The possibilities are numerous and significant. In the years to come, the City of Mission, the Kwantlen, Katzie, Matsqui, and Leq'á:mel First Nations, and their neighbours and partners can find ways to protect ecosystems and cultural values in light of the challenges to come—climate change, population growth, and changes in how people think about forests.

10

A SHARED TERRITORY

CHRIS GRUENWALD KEEPS a flowchart of the projects he plans to tackle as Mission's newest director of forestry. Top of the list is long-term planning. Gruenwald thinks the department could be planning harvests and replanting over ten- or twenty-year periods, rather than on the current five-year cycles. The longer time horizon would allow Mission, Indigenous partners, and other stakeholders to work towards shared goals in a more holistic way and over larger areas of the forest at once.

Like Gruenwald, each of the previous tree farm managers championed new ideas of what the community's forest could be for Mission. Rocky Rockwell emphasized sustained yield, working to improve the condition of the

PRECEDING Rocky Point Campground. *Jason Brawn*

FACING Saunders logging road in autumn. *Kelly Cameron*

forest itself. Bruce Webster navigated the tumultuous recession years but never lost sight of the tree farm's responsibility to provide work to struggling loggers and seasonal labourers. Kim Allan assumed the directorship with a goal of building stronger support for the municipal forest by using revenues and the land itself to develop community assets. Bob O'Neal spearheaded Stave West, a collaborative, multi-stakeholder planning process for the lands on the west side of the tree farm.

The values ascribed to the lands and waters of the Mission forest continue to multiply. The more, the merrier, according to Gruenwald, who sees managing for different values as an essential part of community forestry. Population growth, changes brought on by global warming and other environmental concerns, and the need to manage for cultural assets will require new approaches to the core operations of harvesting and planting. It will also entail building on Mission's current strengths—most importantly, its people.

Bob O'Neal (L), Mission's forestry manager (1990–2013) and director of forestry (2013–19), and Chris Gruenwald (R), Mission's newest director of forestry (2019–). *Jason Brawn*

Growing Populations, New Directions

OF THE SIX people to manage the tree farm, Gruenwald is the first to hail from Mission. He started his forestry education at BCIT, before finishing his degree out east, at the University of New Brunswick. After school, he moved back to Mission and established a forestry consulting firm, Cascadia Environmental Services, with a friend who was an experienced forest technologist. The company specialized in working on area-based tenures—community forests, woodlots, and First Nations woodland licences. The Kwantlen and later the Katzie First Nations employed the firm to help develop their licence applications and management plans.

Getting hired on to the tree farm fulfilled a lifelong goal for Gruenwald, and the job allowed him and his family to stay rooted in the town. "For me especially, one of the things that I cherish about Mission, it's a close-knit community," he said. Gruenwald plans to tap into these relationships and a strong sense of place to build even more connections between the tree farm and the town. This, he'd argue, grows local support for and a better appreciation of the forest's operations.

Some of Mission's small-town feel has waned in recent years as thousands more people have moved into the area. Tens of thousands more are expected to follow in the next two decades. A growing population creates additional challenges for the tree farm. Mission's newer residents are less likely to have economic ties or family histories in forestry, or to view the tree farm as a working forest. They are more likely to look along the shoreline of Stave Lake or up Red Mountain and see recreational spaces with scenic value. Gruenwald expects that, as the community grows, so, too, will pushback against harvesting.

This differentiates community forests from corporate-run operations. Meaningful engagements with local residents and organizations are central to decision-making processes. "We're answerable to the community," Gruenwald noted. "We certainly don't operate in a silo."

He added, "People often look at community engagement as an obligation. I don't really see it that way. Building a community forest is building those relationships."

Throughout the Stave West planning process, Mission has cultivated new partnerships with Indigenous Nations, recreation organizations, and educational institutions. "We know that the area is going to grow in leaps and bounds," said Gruenwald's predecessor O'Neal. "We're going to have the challenges of more public liability [such as] people getting lost on trails. There's going to have to be a real awareness of how to address those challenges."

O'Neal added, "The challenges and the opportunities with the continued public growth of the area will be something we need to work on together." This ethos is modelled in the composition of the Stave West Leadership Team, whose membership includes representatives of the City of Mission, the Kwantlen, Matsqui, and Leq'á:mel First Nations, BC Hydro, and the RCMP. Through ongoing stakeholder engagement and consultation, the team has identified strategies to address the complex issues around multiple uses and safe public access.

Growing interest in the tree farm will translate into new opportunities for economic development in Stave West. Campground development and trail planning is a start. The leadership team has also considered opportunities for water sports, elevated canopy walks, adventure tourism, and other amenities that will attract tourism dollars and new investments into infrastructure and facilities.

Meaningful engage-
ments with local residents
and organizations
**are central to decision-
making processes.**

Hosts Richard and Lynne Lanyon welcome visitors to Kearsley Creek Campground, one of four in Stave West constructed in a six-year period. *Jason Brawn*

On the east side of the tree farm, where a larger number of private properties butt up against more of the forest's edge, the potential for large-scale commercial recreation is limited. But the attributes and values that area residents are likely to associate with the eastern sections—green space, viewpoints, access to trails, and a desire to see wildlife—are the same. None of these values are inherently incompatible with the type and scale of logging carried out in Mission. But it cannot be assumed that the public knows this proposition or agrees with it.

Mission's Forestry Department will need to find ways to balance the shifting priorities of residents with legal and economic targets for harvesting. The job ahead will not be easy. That's because the tree farm is home to a maturing forest and more marketable timber. In 2020, the Ministry of Forests, Lands, Natural Resource Operations and Rural Development adjusted Mission's annual allowable cut to sixty thousand cubic metres per year, a 33 percent increase over the previous five years. Making space for the competing demands on the tree farm will require as much precision as it will vision.

Environmental Change and Fire Risks

THE SUREST THING that can be said about the impacts of climate change on the Mission Tree Farm is that no one yet knows with any precision what they will be. But the general trends are already evident. Weather in southwestern BC has become more unpredictable and frequently more extreme. Summer lasts longer and runs hotter and drier. Winds blow stronger and with increasing frequency. Winter snows are wetter and heavier—or some years do not arrive at all. The region's trademark drizzly springtime has given way to significantly heavier and more intense rainfalls.

What this means for coastal forest management—what will survive, what should be planted—is not entirely clear yet. Some species, especially deciduous ones, will thrive, but a changing climate puts Douglas fir at greater risk for root diseases and needle blight. Drought stresses native species like western red cedar and creates fertile ground for the spread of certain invasive species. Without concerted silviculture efforts, yellow cedar will likely disappear from the tree farm altogether. At ground level, brushy plants grow more thickly and rapidly. Steps to address these concerns need to be weighed against the habitat needs for amphibians, birds, and other wildlife in the forest.

Fires poses the greatest risk. A large-scale stand-clearing fire in Mission's forest would be a threat to life and property and could destroy significant volumes of marketable wood. In BC's Interior, mountain pine beetle and landscape-sized forest fires have frequently wiped out timber supplies, including forests managed under community forest agreements. The province has responded by increasing allowable salvage cutting and granting licence to additional forest areas. Should fire consume large areas of Mission's timber supply, in contrast, there are limited options to increase access to forests elsewhere. Mission is surrounded on three sides by urban development, other area-based tenures, a provincial park, and extremely rugged and high-elevation lands.

Climate change exacerbates the risk and scale of fire, whether it is caused by human activity or by lightning. There are no easy solutions. More people on the tree farm results in more campfire activity, even after campfire bans are put in place. Yet forest managers rely on backcountry users to be their eyes and ears, helping to report illegal or uncontrolled fire activity. And the growing number of groups recreating in Stave West deter those who would dump and torch cars along the Florence Lake Forest Service Road.

If fire were to break out, those living in interface zones would face the greatest danger. For years, Mission's forestry staff have collaborated with local fire services and property owners in Steelhead and Stave Falls to mitigate fire risk. These efforts have included educational outreach on creating defensible zones around built structures to integrating more fire breaks along the tree farm's boundaries.

Harvesting operations add to the risk of fires, through equipment usage and the accumulation of logging waste. Mission already curtails logging activity during extreme drought to prevent the sparking of accidental fires. The build-up of slash debris from logging—the branches and treetops and tree stumps that have no marketable value—is far harder to deal with. Contractors on the tree farm are responsible for reducing the volumes of waste they produce, but there are few commercially attractive options to remove the waste from the landscape.

Summer lasts longer and runs hotter and drier. Winter snows are wetter and heavier—or some years do not arrive at all.

In prior decades, Mission used broadcast burning to deal with waste management on the tree farm. Today, doing so would worsen the air quality in the Fraser Valley and invite significant public and political opposition. Without companies interested in buying the wood waste (for use in pellet production, for example), slash will build up over time. Mission's forest technologist Kelly Cameron has creatively and successfully repurposed some of this wood waste into habitat piles for small mammals, frogs, and other wildlife.

Mission's tree farm managers will have to tackle fire risk from the root up, literally. Forestry staff can choose to replant fewer and more drought-tolerant trees, for instance, and tolerate more deciduous trees in the mix. "The mix is really important," according to O'Neal. "I'm not saying we allow deciduous to overtake our stands, but a certain amount of deciduous is good for us. Big broadleaf maples are good for birds [and] take up plantable space, but they also provide nutrient cycling." Deciduous trees, and specifically maples, are also an important part of the cultural heritage of the Stave Valley.

Finding economically viable strategies to reduce the risk of large fires will require partnering with adjoining tenures on response and mitigation strategies. Two of these tenures are held by the Kwantlen and Katzie Nations. These licensees face the same wood waste problem as Mission. In 2019, the District also worked with a small Fraser Valley company to explore options to process waste on-site. In the long term, however, the Nations and Mission may collectively have more success in finding a buyer for larger volumes of wood debris than any one tenure can produce on its own. Collaboration generates other benefits as well. Working together to plan fire breaks, species mix, fire response training, and emergency protocols is likely to prove more effective when it is done over a larger area, from Alouette to Stave Lake.

Among the challenges facing the tree farm is wood-waste management; working with neighbouring tenures might provide the most sustainable path. *Michelle Rhodes*

Thinking Generationally

THE TREE FARM licence has served Mission's purposes for more than six decades, allowing the District to invest in and manage its local forest base with community needs in mind. Early successes could be chalked up, at least in part, to a special arrangement whereby Mission paid royalties instead of stumpage for its first twenty years. The reduced financial pressures allowed the District to invest revenues into expanding road networks, restoring forest health, and developing expertise in silviculture. Since 1979, however, Mission has been subject to the same rules and stumpage fee as other tree farm licensees.

Under a tree farm licence arrangement, Mission's operation evolved into a community forest with a fully developed system of access and logging roads, mature second-growth timber, and a growing trail and campground network. At any time, a staff of a half-dozen experienced forestry professionals are handling everything from day-to-day operational needs to large-scale projects like the Stave West Forest and Recreation Area.

But the tree farm licence is a blunt-force instrument that was never designed with the needs of local communities in mind. The provincial government first adopted this type of licence after World War II to ensure that companies could sustainably produce high volumes of timber. For smaller tenure holders like Mission, the TFL lacks the flexibility needed to juggle the diverse and at times conflicting demands of managing a community forest.

The District along with leaders from the Matsqui, Leq’á:mel, and Kwantlen Nations **have started on the road towards improving relations and collaborative management.**

Almost all of BC's locally managed forests instead operate under CFAs, or community forest agreements. Mission has weighed the merits of staying as a TFL versus converting to a CFA. Such a move would provide certain advantages, including more preferential stumpage rates and greater ease in managing cutting permits. Under a CFA, the city could more easily protect non-market values, such as by pulling selected forest habitat out of production for conservation purposes.

Unlike CFA holders, however, Mission has already made decades of investments into a well-developed road network, forest restoration, and silviculture. Transitioning to a CFA may provide fewer financial advantages than it would for other communities, and the move would entail changes to the management of the tenure that may or may not be supported politically. The tree farm licence may not have been designed for small communities like Mission, but it is a structure to which the municipality has adapted well over the years.

• • •

The District along with leaders from the Matsqui, Leq'á:mel, and Kwantlen First Nations have started on the road towards improving relations and collaborative management. The development of the SWFRA has evolved into an inclusive information-sharing and land-use decision-making process, with tangible results. Following the adoption of the *Stave West Master Plan*, the multi-stakeholder leadership team has overseen the creation of new recreation amenities, hosted K–12 and post-secondary tours and field schools, and tackled parking, road use, and security issues along the Florence Lake Forest Service Road.

The next twenty to fifty years will further test the resilience of the Mission's community forest operation. The pace of social and ecological change in the forest is quickening. Each year, the tree farm attracts more people from across southwestern BC. The core ingredients that were instrumental in Mission's past successes—leadership, investment, adaptation, collaboration—will be even more essential in the years to come.

Resiliency requires an additional element: *thinking generationally*. Generational thinking means approaching big-picture questions with an eye towards the long term. This goes beyond the next five, ten, or even twenty years, although these are good places to start. Generational thinkers evaluate decisions—as Kwantlen, Leq'á:mel, and Matsqui Elders do—on how they will impact the health and livelihoods of the several generations that follow in their footsteps.

The name given to Stave West, mekw'wa't a'xwest ikw'elo' (everyone shares here), reflects generational thinking. It honours those who once lived and worked this place. It respects the generations yet to be born.

The trees planted today will outlive the people who planted them. The choices that are made about what to plant and harvest, the values that are placed on nature as commodity or cultural resource, the ways in which what to use is balanced with what to protect—these are decisions that last beyond the lifetime of any one forest manager, backcountry camper, or local resident.

Generational thinking is an innate feature of community forestry. Working with Indigenous partners, political leaders, and community volunteers allows forest managers and everyone else involved to take stock of the values that are important to their communities. Local communities are the ones who will have to figure out how to adapt to changes resulting from climate change and population growth. They are the ones most impacted by the trade-offs made between the economic needs of today and the forests of tomorrow.

The evolution of the Mission Tree Farm is a story of taking risks but also of making smart investments that produce long-term benefits to its community. The future will look different. Forest management will be more collaborative than in past years and, critically, involve sharing benefits more widely. And while Mission's history of community forest management dates back a few generations, longer than in any other town in Canada, the whole enterprise is still younger than many of the trees rooted here. Mission and its tree farm still have a lot of growing left to do.

Playing along the shore of Kearsley Creek Campground. The next generation of visitors will help shape the future story of Canada's oldest community forest. *Jason Brawn*

ACKNOWLEDGEMENTS

As is fitting for a book detailing the evolution of a community's forest, *The Tree Farm* was, from start to finish, a collective effort. This project would not have happened without the incredible commitment, insight, and patience of Bob O'Neal, the former director of forestry for the District of Mission. Bob's passion for telling the story of the tree farm and of Stave West is infectious, yet always tempered with a healthy dose of realism about the challenges that lay ahead. The idea for the book originated with members of the Mission Forestry Department, as a way of commemorating the sixtieth anniversary of the tree farm. Bob was supportive of different approaches to the putting the book together, and as it became evident that the sixtieth was going to come and go before the manuscript was done, Bob and his successor, Chris Gruenwald, were also quite forgiving of the extra time needed for revisions.

Both Bob and Chris wrangled budgets for the project, shared their time and perspectives during hours of interviews and other meetings, and organized feedback on the various iterations of the book as it was in development. *The Tree Farm* was funded by the City of Mission and its Forestry Department, and the book would not have happened if not for the municipality's faith in the value of sharing the story of its community forest. Keeping the project moving forward, Terry Hood of North Shore Project Leadership lent his time and his attention at every stage of book's compilation, including organizing planning meetings, scheduling photo shoots, recommending potential publishers, and sharing his experiences as a consultant on the Stave West.

I cannot thank enough as well the time and feedback shared by Tumia Knott and Drew Atkins of the Kwantlen First Nation; Phil Sherwood, Jason Thompson, and Shawn Gabriel of the Leq'á:mel First Nation; and Cindy Collins of the Matsqui First Nation. Tumia, who always has a million things on her plate, nonetheless gave up several hours one sunny afternoon with me at the səyeḿ qwantlen offices in Fort Langley. She provided her impressions of the manuscript's first draft and patiently filled in crucial gaps in my knowledge of how the Kwantlen's relationship with Mission has changed over time. Jason and Shawn likewise sat down with Chris and me to discuss the third draft and to raise questions and concerns about the telling of the history of the Stave Valley. I hope the version of the book that went to press captures needed changes, although I know it remains an imperfect and incomplete telling of these long and complex cultural histories and relationships.

Members of Mission's Forestry Department contributed to the book project in different ways, most notably forest technologist Kelly Cameron. In addition to having a wealth of knowledge and experience from more than twenty-seven years of working on the tree farm, Kelly is also a keen and skilled photographer. Many of her photos are found in the book. On several occasions, foreman Brad Laughlin was drafted into driving me and photographers to active harvesting sites and other spots on the tree farm, at one point getting a flat tire well up on Hoover Lake Road, and at the end of the workday no less. As valuable as Brad's time was, I especially appreciated how patient he was while I peppered him with questions about the tree farm.

In sitting down with Kim Allan as well as Bob and Chris, I was able to hear and compare the perspectives of the three people who have spent a combined three decades overseeing TFL 26. Kim's input adds a critical piece to the book's later chapters. Kim's predecessor, Bruce Webster, passed away some years back. In his stead, Bruce's son, Graham Webster—also a forester—told his story to me over the phone. He also sent me physical copies of photographs taken of Bruce as a young man and after he retired. By the end of our conversation, I realized that if there was only one person in the history of the tree farm operations that I could have had the opportunity to have met, it would have been Bruce.

Archival resources, notably back issues of the *Fraser Valley Record* and photographs of Mission's forestry history, were accessed through Mission Community Archives. Mission's archivist, Val Billesberger, pulled materials, scanned photographs, and provided input on additional historical resources worth reviewing. Val also allowed me to carefully work through a collection of photographs that had been recently donated by Howard "Smokey" Murdoch's family, despite the collection not yet having been appraised or accessioned. The photographs form a critical part of the story of the tree farm.

Additional historical research and support was provided by Jillian Wheatley, a BCIT practicum student who, in 2017, helped with archival searches and conducted oral history interviews. Many thanks as well go to Jillian's interviewees—Norm Horn, George Kocsis, and Lorne Rockwell. Some of their stories and recollections are included in the book, and these recorded histories will be valued resources for future researchers of Mission's forest past. In addition, Mission city councillor Ken Herar provided wonderful source material on Naranjan Grewal and other prominent South Asian Canadian business leaders in Mission from the 1940s and 1950s. Grewal's fascinating story warrants its own book someday.

For the development of mountain biking culture and trails in the Mission forest, I turned to Rocky Blondin and Frank Gunderman to help fill in the details. Rocky is not only an active promoter of the sport in Mission, he has also made important contributions to the Stave West planning process. I'd also like to thank Rose Schroeder of the Backcountry Horsemen of BC, Yarrow chapter. I was on a tour of the Kearsley Creek Campground before it first opened and talked with Rose at the time about the work the organization had done in building corrals at the site. This was one of those right-place, right-time chances to chat, rather than a formal interview for the book, but her insights helped inform my understanding of the needs of recreational horseback riders.

Other past and current members of Mission's Forestry Department provided informational and logistical support. Prior to his retirement in 2019, I also interviewed Dave Heyes, then manager of forest business, who introduced me to the history of "evergreen contracts" and detailed impacts from anticipated changes in stumpage. Michelle Weisgerber and Erika Duplisse both provided clerical support, and Michelle provided some research assistance to boot.

I want to extend my gratitude as well to the past and current members of the Stave West Leadership Team, including BC MLA Pam Alexis, who was busy with her then mayoral campaign activities at the time of our interview. In the process of writing this book, I sat in on several SWLT meetings, including crucial recreation planning presentations. This time allowed me to see how the folks at the table worked on shared concerns. Prior to becoming chief administrative officer for Fernie, BC, Michael Boronowski organized these meetings. Michael was manager of civic engagement and corporate initiatives for the District of Mission and, like Terry, kept business moving in a forward direction. Michael helped with logistical support for the book. Claire Seeley, Mission's manager of tourism, has graciously and capably stepped into this role of late.

Prior to starting this book, I knew very little about mosses, ferns, and fungi, certainly not enough to write about what species I might find along the Roy Kittles Trail. University of the Fraser Valley geography students and naturalists-in-training Donovan Toews and Michelle Lefebvre joined me out on the trail one morning to work on plant observations and photography. They were the ones to identify the many species of moss seen along the trail included in chapter two and one of Michelle's photos is found in the book as well.

Many people captured the beauty and features of the Mission Tree Farm in photography and drawing. Jason Brawn's photographs are found throughout, including on the cover and front pages. Jason also sent his drone up for pictures of the Stave delta, Ruskin Dam, and Foreshore Flats. Some of Kelly Cameron's photos were of spots and trees that only someone with extensive knowledge of the tree farm would know to find. Additional photographs were taken by students enrolled in the UFV's graphic and digital design (GDD) program, under the direction of Alex Wells. Justine Robinson, a UFV student in visual arts, spent the day on a photoshoot for the book, and another UFV student, Emily Gauthier, created the lovely plant sketches placed throughout the book. Other recent photos were taken by Terry Hood, UFV faculty member Karin Jager, and me.

Karin Jager and Jen Deon, UFV GDD faculty, helped organize student participation in the book project. Among the projects Karin and Jen oversaw was a class project in which GDD students designed mock-ups of the book cover and inside page for *The Tree Farm*. Bob, Terry, Drew, Tumia, and Michelle all joined in for the presentation of the final projects in late November 2017. That was the first point at which some of us could imagine what this book would look like on someone's coffee table.

Finally, our collective thank you on behalf of all of us in Mission goes out to Page Two, the publishing services firm that managed the design, editing, and printing details of the book. Unforeseen circumstances meant that the final revisions for this book took far longer than anticipated to finish, and the team at Page Two was incredibly patient as this work was completed.

NOTES

Introduction: A Watershed Moment

... *"a wet rag on a salad"*... Tom Robbins, *Another Roadside Attraction* (Bantam, 1990, 1971), 221.

... *Tree Farm Licence (TFL) 26 has expanded multiple times*... "Tree Farm Project Seen as Boost to Economy," *Fraser Valley Record (FVR)*, August 26, 1964.

1: Seeds of an Idea

Where people still practise strong cultural traditions... Benjamin D. Hodgdon, Francisco Chapela, and David B. Bray, *Mexican Community Forestry: Enterprises and Associations as a Response to Barriers* (Centre for People and Forests and Rainforest Alliance, 2013), 2.

Day-to-day management is left to forestry professionals... Ryan C.L. Bullock and Kevin Hanna, *Community Forestry: Local Values, Conflict, and Forest Governance* (New York: Cambridge University Press, 2012), 148, 152.

The Municipality of North Cowichan on Vancouver Island and the District of Mission established municipal forest reserves two years apart... Kim Allan and Darrell Frank, "Community Forests in British Columbia: Models That Work," *Forestry Chronicle* 70, no. 6 (1994), 723.

... *the Village of Hope had suggested as much*... "Municipal Tree Farm Suggested," *Chilliwack Progress*, June 18, 1958, sec. 1, p. 5.

Residents in the Slocan Valley and Smithers-Kispiox Valley also pushed... Jeremy Wilson, *Talk and Log: Wilderness Politics in British Columbia, 1965–96* (Vancouver: UBC Press), 144.

Kaslo and Creston did eventually procure... British Columbia Community Forest Association, "A Brief History of Community Forestry in BC," bccfa.ca.

... *more than 90 percent of forest lands are held by the Crown*... More than 95 percent of BC's forest land is held by the Crown. Government of British Columbia, "Timber Tenures in British Columbia," 2. For more on the Revelstoke Community Forest, see

Ella Furness, Howard Harshaw, and Harry Nelson, "Community Forestry in British Columbia: Policy Progression and Public Participation," *Forest Policy and Economics* 58 (2015).

... *the Fordist era* ... Roger Hayter, *Flexible Crossroads: The Restructuring of British Columbia's Forest Economy* (Vancouver: UBC Press, 2000), chapter 2.

In the 1990s, the province loosened requirements ... Christopher Lyon and John R. Parkins, "Toward a Social Theory of Resilience: Social Systems, Cultural Systems, and Collective Action in Transitioning Forest-Based Communities," *Rural Sociology* 78, no. 4 (2013), 538.

Small cities and regional governments lobbied for greater control ... Kirsten McIlveen and Michelle Rhodes, "Community Forestry in an Age of Crisis," in *Community Forestry in Canada: Lessons from Policy and Practice*, ed. Sara Teitelbaum (UBC Press, 2016), 187.

... *a CFA is a renewable, area-based licence* ... Furness, Harshaw, and Nelson, "Community Forestry in British Columbia," 87; "A Brief History of Community Forestry in BC," bccfa.ca.

... *more than five dozen distinct community forest operations* ... BC Community Forest Association (May 2020), bccfa.ca.

Still other CFAs are overseen ... Chinook Community Forest, chinookcomfor.ca; Cheakamus Community Forest, cheakamuscommunityforest.com.

2: The Stave Valley

... *better known as the Ice Age* ... Robinson Meyer, "Geology's Timekeepers are Feuding," *Atlantic* (July 20, 2018).

This glacial event reached its maximum ... John Menzies, *Modern and Past Glacial Environments* (Elsevier, 2002), 42.

In British Columbia, glaciers developed in alpine regions ... Mike Church and J.M. Ryder, "Physiography of British Columbia," in *Compendium of Forest Hydrology and Geomorphology in British Columbia*, vol. 1 (Victoria, BC: BC Ministry for Forest and Range, Forest Science Program, 2010), 29; Bryan J. Mood and Dan Smith, "Holocene Glacier Activity in the British Columbia Coast Mountains, Canada," *Quaternary Science Reviews* 128 (2015), 15.

... *glaciers reached their southwestern and westernmost extent by 16,500 YBP* ... Brian Menounes, Gerald Osborn, John J. Clague, and Brian Luckman, "Late Pleistocene and Holocene Glacier Fluctuations in Western Canada," *Quaternary Science Reviews* 28 (2009), 2051.

... *ice covered the Stave Valley more than two thousand metres deep* ... Robert Gilbert and Joseph Desloges, "The Late Quaternary Sedimentary Record of Stave Lake, Southwestern British Columbia," *Canadian Journal of Earth Sciences* 29 (1992), 2002.

... *a phenomenon called isostatic rebound* ... John E. Armstrong, *Environmental and Engineering Applications of the Surficial Geology of the Fraser Lowland, British Columbia*, paper 83-23 (Ottawa: Geologic Society of Canada, 1984), 26.

... *Michael Church wrote* ... Church and Ryder, "Physiography of British Columbia," 26.

... *the U-shaped valleys found in the Coast Mountains* ... David Huntley and David

Thompson, "Fraser Valley Geo-Tour: Bedrock, Glacial Deposits, Recent Sediments, Geologic Hazards and Applied Geology: Sumas Mountain and Abbotsford Area," guide prepared for MineralsEd and Natural Resources Canada, Geologic Survey of Canada (2013).

... this rock was granitic, formed during periods of great volcanic activity... Armstrong, *Environmental and Engineering Applications*, 6–7, 22.

Silt, shales, sands, and similar materials covered the granite bedrock in the lower Stave area. Gilbert and Desloges, "The Late Quaternary Sedimentary Record of Stave Lake"; Menounes et al., "Late Pleistocene and Holocene Glacier Fluctuations," 2051–52; Armstrong, *Environmental and Engineering Applications*, 6–7, 15, 22; Huntley and Thompson, "Fraser Valley Geo-Tour," 7; Morgan Ritchie, Dana Lepofsky, Sue Formosa, Marko Porcic, and Kevan Edinborough, "Beyond Culture History: Coast Salish Settlement Patterning and Demography in the Fraser Valley, BC," *Journal of Anthropological Anthropology* 43 (2016), 144.

The geologic profile of the lower Stave... Armstrong, *Environmental and Engineering Applications*, 6–10.

... peaks in the southern Coast Mountains were free of snow... Mood and Smith, "Holocene Glacier Activity," 15.

Forests, fish, and people closely followed the retreat of the ice... Ritchie et al., "Beyond Culture History," 143.

Permanent settlement occurred in the Stave... Ritchie et al., "Beyond Culture History," 143–44.

... glaciers expanded again from between 6,900 to 5,600 YBP, and their retreat was followed by an increase in the size and number of settlements... Ritchie et al., "Beyond Culture History," 145–46.

... reported finding gold in the vicinity of Clearwater and Glacier Creeks... Charles A. Miller, *"The Golden Mountains": Chronicles of Valley and Coast Mines* (Mission, BC: *Fraser Valley Record*, 1973). Glacier Creek empties into Stave Lake near Glacier Point, just south of Tingle Creek.

... Stave Glacier in southeastern Garibaldi Park... Jill Harvey, Dan Smith, Sarah Laxton, Joseph Desloges, and Sandra Allen, "Mid-Holocene Glacier Expansion Between 7500 and 4000 cal. yr BP in the British Columbia Coast Mountains, Canada," *The Holocene* 22, no. 9 (2012), 983; Johannes Koch, Brian Menounes, and John Clague, "Glacier Change in Garibaldi Provincial Park, Southern Coast Mountains, British Columbia, Since the Little Ice Age," *Global and Planetary Change* 66 (2009), 163–64; Mauri Pelto, "Stave Glacier Retreat, British Columbia," *From a Glaciers Perspective* (blog), December 12, 2012, glacierchange.wordpress.com.

... 150 glaciers in Garibaldi Park... Koch, Menounes, and Clague, "Glacier Change in Garibaldi Provincial Park," 162.

*... the yellow cedar (*Chamaecyparis nootkantensis*) and the western red cedar (*Thuja plicata*)...* John Krapek and Brian Buma, "Yellow Cedar: Climate Change and Natural History at Odds," *Frontiers in Ecology and Environment* 13, no. 5 (2015), 280–81.

... one would have to go back nearly a hundred million years to find a common ancestor... Damon L. Little, "Evolution and Circumscription of the True Cypress (Cupressaceae: *Cupressus*)," *Systematic Biology* 31, no. 3 (2006), 466–67.

Yellow cedar emerged as a distinct species... Randall G. Terry, Matthew I. Pyne, Jim A. Bartel, and Robert P. Adams, "A Molecular Biogeography of the New World Cypress (*Callitropsis, Hesperocyparis; Cupressaceae*)," *Plant Systematics and Evolution* 302 (2016), 923; Krapek and Buma, "Yellow Cedar," 280.

... *the western red cedar's closest relatives are in Korea and Japan*... Jian-Hua Li and Qiao-Ping Xiang, "Phylogeny and Biogeography of *Thuja* L. (Cupressaceae), and Eastern Asian and North American Disjunct Genus," *Journal of Integrative Plant Biology* 47, no. 6 (2005), 651–59.

... *both western red and yellow cedar are relatively new arrivals*... Lisa M. O'Connell, Kermit Ritland, and Stacey Lee Thompson, "Patterns of Post-Glacial colonization by Western Red Cedar (*Thuja plicata*, Cupressaceae) as revealed by microsatellite markers," *Botany* 86, (2008), 199; Church and Ryder, "Physiography of British Columbia," 31.

... *between 1.8 and 2.3 metres of precipitation*... Averages based on 1981–2010 data. Lower totals are for the Mission West Abbey station, and higher totals were recorded at the Stave Falls station. Environment and Climate Change Canada, "1981–2010 Climate Normals & Averages," climate.weather.gc.ca.

The Stó:lō used t'ege'lhp and its berries, t'áqe... When possible, the upriver dialect of Halq'eméylem is used in text for purposes of consistency. Brent Douglas Galloway, *Halq'eméylem-English Stolo Dictionary* (University of the Fraser Valley, 2004), ufv.ca; Brian Compton and Donna Gerdis, *Native Plants, Peoples, and Animals: Halkomelem*, Simon Fraser University, sfu.ca. See also Brent Douglas Galloway, *Dictionary of Upriver Halkomelem* (University of California Publications, Linguistics, v. 141, 2009).

... *types of bats*... Lorraine Andrusiak and Mike Sarell, "Evaluation of Experimental Artificial Rock Roosts for Bats," report prepared for the Fish and Wildlife Conservation Program (Project No: COA-F19-W-2701), March 2019; Juliet Craig, Mike Sarell, and Susan Holroyd, "Got Bats?: BC Community Bat Project Frequently Asked Questions," prepared for the BC Community Bat Project Initiative, 2014.

... *reduced the size and quality of habitat in the Stave Valley*... Fish & Wildlife Compensation Program, Stave River Watershed Action Plan, version 9.0, September 28, 2017; K.L. MacKenzie, "Wildlife Habitat Interpretations of the Terrestrial Ecosystems on Mission Tree Farm License 26," report produced for District of Mission, March 2004; Caroline Astley, "Assessing the Location of Existing Reserve Patches and Wildlife Corridors TFL 26-Mission, BC," report prepared for the District of Mission, 2007.

Human modifications have most dramatically altered lake habitats... John Stockner and Darren Bos, "History of Wahleach and Stave Reservoirs: A Paleolimnological Perspective," report prepared for Parks Canada, 2002.

... *the reintroduction of the Roosevelt elk*... *Lower Mainland Roosevelt Elk Recovery Project (LMRERP): Stave Lake Watershed Release (2007–2008)* (BC Ministry of Environment, Wildlife Branch, June 2008).

3: Roots

... *shinglewood*... US Forest Service, *An American Wood* (Washington, DC: US GPO, 1937).

... *cypress*... E-Flora BC, Electronic atlas of the Flora of British Columbia, ibis.geog.ubc.ca/biodiversity/eflora; District of Mission, *Annual Reports* for years 1960–1970, unpublished.

... pá:yelhp in Halq'eméylem, xhpey'ulhp n Halkomelem... Compton and Gerdis, *Native Peoples, Plants and Animals: Halkomelem*.

You know that a long time ago there was a very generous man... Albert (Sonny) McHalsie (Naxaxahlts'i), "We Have to Take Care of Everything That Belongs to Us," in *Be of Good Mind: Essays on the Coast Salish*, ed. Bruce Granville Miller (Vancouver: UBC Press, 2008), 104–105.

... *Stó:lō harvesters gave thanks through prayer*... Naxaxahlts'i, "We Have to Take Care of Everything," 105.

They used cedar wood... Naxaxahlts'i, "We Have to Take Care of Everything," 105; Jim Pojar and Andy MacKinnon, *Plants of the Pacific Northwest Coast*, rev. ed. (Vancouver: Lone Pine Publishing, 2004), 41.

... *the softer inner bark of pá:yelhp*... Compton and Gerdis, *Native Peoples, Plants and Animals: Halkomelem*.

The wetlands were reconstructed... Fraser Valley Watersheds Coalition, *Stave River Watershed—Restoring Salmon Rearing and Overwintering Habitat* (Site 2–Phase 1), report prepared for Fish and Wildlife Conservation Program, 2017, gov.bc.ca.

Radiocarbon dating... Ritchie et al. "Beyond Culture History," 143–45; Duncan McLaren, "The Occupational History of the Stave Watershed," in *Archaeology of the Lower Fraser River Region* (Burnaby, BC: Simon Fraser University, 2017), 150.

Artifacts left over from settlements and campfires... McLaren, "Occupational History of the Stave Watershed," 151–52.

Stó:lō hunters and harvesters also set fires... Jeff Oliver, *Landscapes and Social Transformations on the Northwest Coast: Colonial Encounters in the Fraser Valley* (Tucson, AZ: University of Arizona Press), 35–36.

Hunters employed pack dogs... Mike Rousseau, "A Culture Historic Synthesis and Changes in Human Mobility, Sedentism, Subsistence, Settlement, and Population on the Canadian Plateau, 7000–200 B.P.," in *Complex Hunter-Gatherers: Evolution and Organization of Prehistoric Communities on the Plateau of Northwestern North America*, ed. William C. Prentiss and Ian Kujit (University of Utah Press, 2004), 3–22.

... *an entire community or settlement owned the resource or land* Keith Thor Carlson, *The Power of Place, The Problem of Time* (Toronto: University of Toronto Press, 2010), 47–48.

Roots and rushes were collected... Natasha Lyons, Tanja Hoffman, Debbie Miller, Stephanie Huddlestan, Roma Leon, and Kelly Squires, "Katzie and Wapato: An Archaeological Love Story," *Archaeologies* 14, no. 1 (2018), 7–29.

Cedar met hundreds of the day-to-day needs Nancy Turner, *Plants in British Columbia Indian Technology*, BC Provincial Museum Handbook No. 38 (Victoria: BC Provincial Museum, 1979), 74–89.

The Stó:lō fashioned fir branches into harpoons... Turner, *Plants in British Columbia Indian Technology*, 112.

Boiled hemlock bark... Turner, *Plants in British Columbia Indian Technology*, 115; Compton and Gerdis, *Native Peoples, Plants, and Animals: Halkomelem.*

... sturgeon, that grew "as long as canoes." Oliver, *Landscapes and Social Transformations on the Northwest Coast*, 69.

How the ravens arrived in the Stave... This story is captured by Keith Thor Carlson, who draws from Old Pierre's telling of this story as well as additional, albeit limited, oral histories recorded in the late nineteenth and early twentieth centuries. These are found in Carlson's *The Problem of Place, the Problem of Time* (Toronto: University of Toronto Press, 2010), 100–106. The Katzie Nation today trace their origins back to the Pitt Lake settlement; the Kwantlen, to the foot of the Alouette River and surrounding region.

... sxayə'qs, or "everyone's landing place." Kevin Neary, *Cultural Heritage Sites Literature Review: Kwantlen Territory Knowledge Project*, report prepared for Kwantlen First Nation by Traditions Consulting Services, 2011, 58.

... a small village of twenty-two residents along the lower Stave Neary, *Cultural Heritage Sites Literature Review*, 14–15.

... the explorer George Vancouver noted... Cole Harris, *The Resettlement of British Columbia: Essays on Colonialism and Geographical Change* (Vancouver: UBC Press, 1997), 4, 12; Carlson, *The Problem of Place, the Problem of Time*, 91–96.

... Simon Fraser wrote in his journals... Robert Boyd, "Commentary on Early Contact-Era Smallpox in the Pacific Northwest," *Ethnohistory* 43, no. 2 (1996), 320.

... "If you dig today on the site of any of the old villages..." Harris, *Resettlement of British Columbia*, 8–9.

Old Pierre's son, Simon, shared the location... Carlson, *The Problem of Place, the Problem of Time*, 97.

... Sxwoxwiymelh, or "a lot of people died at once." Harris, *Resettlement of British Columbia*, 9.

The Stó:lō abandoned smaller villages for more central places. Carlson, *The Problem of Place, the Problem of Time*, 98.

... found themselves joining with the Katzie and Kwantlen. Carlson, *The Problem of Place, the Problem of Time*, 101–104.

By the 1820s, hundreds of people lived in the Kwantlen villages. Harris, *Resettlement of British Columbia*, 69.

After the arrival of the Hudson's Bay Company near Fort Langley... Neary, *Cultural Heritage Sites Literature Review*, 13; Carlson, *The Problem of Place, the Problem of Time*, 106.

The reserve system imposed forced sedentarism... Neary, *Cultural Heritage Sites Literature Review*, 7.

4: Transforming the Stave

... the Stave River drops eighty metres... Meg Stanley and Hugh Wilson, *Station Normal: The Power of the Stave River* (Vancouver: Douglas & McIntyre, 2001), 17.

Vancouver's appetite for power... Mary Taylor, *Development of the Electricity Industry in British Columbia* (unpublished master's thesis, UBC, 1965), 20.

... the completion of the Ruskin Dam... Matthew Evenden, *Fish versus Power: An Environmental History of the Fraser River* (Cambridge, 2004), 68.

... *British Columbia Timber Lands and Electric Company*... Stanley and Wilson, *Station Normal*, 20.

... *Stave River Electric and Power Company*... British Columbia, "Bank of B.C. Against Stave River Co. Bill," *Journals of the Legislative Assembly of the Province of British Columbia*, vol. 24 (Victoria: Queen's Printer, 1895).

... *Stave Lake Power Company*... Stanley and Wilson note that the SLPC formed in 1899 and received provincial approval to generate electricity at the site in 1900. Evenden provides two different dates, with the company having formed in 1901, after having obtained rights even earlier, in 1897. I have chosen to use the dates in Stanley and Wilson, as they appear to have drawn these dates directly from primary documentary sources. (Stanley and Wilson, *Station Normal*, 23; Evenden, *Fish versus Power*, 60.)

... *the SLPC laid out cash*... Stanley and Wilson, *Station Normal*, 26.

... *Western Canadian Power Company*... George Robert Graham Conway, *Province of British Columbia* (Government of Canada, Dominion Water Power Branch, 1915), 59.

The first power was delivered to Vancouver... British Columbia, Dept. of Lands and Forests, *Water Powers: British Columbia* (Victoria: Queen's Press, 1930), 37; Stanley and Wilson, *Station Normal*, 29, 50.

The Stave Falls generating station did not produce for the town of Mission... "30-Year Wait Rewarded: BC Electric Power Line from Falls Finally Reaches District Community," *FVR*, December 9, 1949, sec. 1, p. 1.

... *the British Columbia Electric Railway (BCER) company bought out the WCPC.* Stanley and Wilson, *Station Normal*, 26.

BCER invested heavily in the Stave project... Stanley and Wilson, *Station Normal*, 51.

... *the BCER needed to double production*... Evenden, *Fish versus Power*, 64; BC Hydro, "Bridge River History and Reservoir Information," 2018, bchydro.com.

The BCER quickly erected the Ruskin Dam... Evenden, *Fish versus Power*, 67.

The dam created a new lake behind it... Stanley and Wilson, *Station Normal*, 51.

... *in a time before environmental assessments.* Evenden, *Fish versus Power*, 76.

... *Mission's Rod and Gun Club repeatedly advocated for fish ladders*... "Young Trout Teem at Dam, Unable to Ascend Stave," *FVR*, August 7, 1947, sec. 1, p. 10.

Both lakes today are classed as ultra-oligotrophic. T.B. Stables and C.J. Perrin, "Stave River Water Use Plan—Fish Biomass Assessment," project SFLMON#3, report prepared for BC Hydro, January 31, 2016, bchydro.com; J. Bruce and J. Beer, "Stave River Water Use Plan—Pelagic Monitor (Nutrient Load / Total Carbon Levels)," project SFLMON-01, report prepared for BC Hydro, March 2017, bchydro.com.

The Kwantlen and Katzie First Nations have successfully pushed for greater say... Rafferty Baker, "BC Hydro Pressured to fund New Salmon Infrastructure in Alouette Watershed," BC News, September 28, 2018, cbc.ca; BC Hydro, *Stave River Water Use Plan*, 2013, bchydro.com.

The first loggers... Arnold M. McCombs and Wilfrid W. Chittenden, *The Fraser Valley Challenge: An Illustrated Account of Logging and Sawmilling in the Fraser Valley* (Harrison Hot Springs, BC: Treeline Publishing, 1990), 55.

These mills were followed by others... McCombs and Chittenden, *Fraser Valley Challenge*, 21.

... *an evolving and generous land grant and leasing program.* Jean Barman, "Beyond Chinatown: Chinese Men and Indigenous Women in Early British Columbia," *BC Studies* 177 (2013), 127.

... *Fraser River Saw Mills Ltd., acquired the rights to log*... McCombs and Chittenden, *Fraser Valley Challenge*, 31; Government of BC, *Public Inquiries Act—Report of the Commissioner relating to The Forest Resources of British Columbia*, The Honorable Gordon McG. Sloan, Commissioner (Victoria, BC, 1945), 89–92.

... *transformed this struggling company into Fraser Mills*... McCombs and Chittenden, *Fraser Valley Challenge*, 31; Taylor, *Development of the Electricity Industry*, 42.

One cedar might measure upwards of three metres in diameter... Barman, "Beyond Chinatown," 127.

... *men worked in teams*... "Life at Stave Falls Far Different 60 Years ago, Pioneers Recall," *FVR*, February 6, 1947, 1.

Edward H. Heaps had assumed control of the Lobb Shingle Mill at Ruskin... McCombs and Chittenden, *Fraser Valley Challenge*, 58.

When he arrived in Vancouver, Heaps purchased a pair of shingle machines... Robert Griffin, "The Shingle Sawing Machine in British Columbia, 1901 to 1925," *Material Culture Review* 12 (1981), 30.

... *Keystone Logging Company*... McCombs and Chittenden, *Fraser Valley Challenge*, 60; "Economy Based Largely on Products of Forests," *FVR*, October 14, 1964, sec. 2, p. 7.

... *Abernethy Lougheed Logging Co*.... McCombs and Chittenden, *Fraser Valley Challenge*, 59.

... *the number of logging railroads built in the Fraser Valley doubled*... Robert D. Turner, "Logging Railroads and Locomotives in British Columbia: A Background Summary and the Preservation Record," *Material Culture Review* 12 (1981), 5.

Rail builders drove tall timbers... Miller, *"The Golden Mountains."*

... *the world's shortest chartered railroad*... Stanley and Wilson, *Station Normal*, 27, 42.

... *the company used the A-frame logging technique*... McCombs and Chittenden, *Fraser Valley Challenge*, 61. A-frame logging involves the construction of a giant A using two very tall timbers, braced near the top with one or more cross-beams. Cables would then be connected to the top of the frame, and a steam donkey or diesel engine would winch the logs attached to the ends of the cables upwards for yarding and loading.

At peak, the company moved more than one million board feet... Stanley and Wilson, *Station Normal*, 42.

... *the railway would soon be replaced*... Taylor, *Development of the Electricity Industry*, 128.

By 1920, truck logging arrived... Taylor, *Development of the Electricity Industry*, 129.

... *the end of railroad logging in the valley*... Stanley and Wilson, *Station Normal*, 43.

BC initially exported much of its unprocessed cedar... Griffin, "Shingle Sawing Machine," 23.

... *"cutting the logs 16 feet long..."* "Life at Stave Falls Far Different 60 Years Ago, Pioneers Recall," *FVR*, February 6, 1947; "Co-op Mill Gave Ruskin Its Start in 1880," *FVR*, October 11, 1967, sec. 4., p. 2.

Alternatively, the log would be further subdivided... Gordon Hak, "Bus Griffiths' *Now You're Logging*: A Graphic Novel about British Columbia Coastal Logging in the 1930s," *Labour / Le Travail* 55 (2005), 180; McCombs and Chittenden, *Fraser Valley Challenge*, 67.

The mills and logging companies employed crews of recent arrivals... Guy Symonds, "Canada's Only Municipal Tree Farm Holds Unlimited Possibilities," *FVR*, June 27, 1968, sec. 1, p. 7; Taylor, *Development of the Electricity Industry*, 111.

... *"they'd pull the boards and the water would sluice down..."* Lorne Rockwell, interview, conducted by Jillian Wheatley at Mission Public Works Building, 2017.

One of the largest flume systems in the Lower Mainland... McCombs and Chittenden, *Fraser Valley Challenge*, 67.
... *A.S. Cameron (later a prominent mill operator in Mission) recalled*... "Logging Feeds World Markets," *FVR*, October 11, 1967, spec. sec., p. 3.
... *cedar bolt cutting and salvage operations provided a bulwark against the worst effects of the Great Depression*... "Economy Based Largely on Products of Forests," *FVR*, October 14, 1964.
... *US federal government worked to prevent competition*... Government of BC, *Public Inquiries Act*, 59.
... Lobb Shingle Mill... McCombs and Chittenden, *Fraser Valley Challenge*, 21; "Life at Stave Falls Far Different 60 Years Ago, Pioneers Recall," *FVR*, February 6, 1947.
[The McTaggart mill] had a switch on the CPR ... "Early Logging Operation Remembered by Pioneer," *FVR*, February 1, 1955, sec. 1, p. 5.
Dozens of additional mills sprang up... McCombs and Chittenden, *Fraser Valley Challenge*, 69–71; "Shingle Mill Being Erected on Riverfront," *FVR*, April 3, 1947, 1.
... *Stave Lake Cedar Mill*... McCombs and Chittenden, *Fraser Valley Challenge*, 59.
... *Abernethy Lougheed Mill*... McCombs and Chittenden, *Fraser Valley Challenge*, 68.
... *recent arrivals from Asia*... "Canada's Only Municipal Tree Farm Holds Unlimited Possibilities," *FVR*, June 12, 1968, sec. 1, p. 7.
... *Japanese and Chinese workers earned far less*... Anne Dore, *From Harbour to Harvest: The Diverse Paths of Japanese-Canadians to Land Ownership, Farming, and the Making of Community in the Fraser Valley, 1904–1942* (master's thesis, Simon Fraser University, 2004), 35.
Just north of the CPR trestle... "Co-op Mill Gave Ruskin Its Start in 1880," *FVR*, October 11, 1967; "Life at Stave Falls Far Different 60 Years Ago, Pioneers Recall," *FVR*, February 6, 1947.
... *more than 2,500 Chinese immigrants*... Barman, "Beyond Chinatown," 43; Hak, "Bus Griffiths' *Now You're Logging*," 177.
More than two thousand first- and second-generation Japanese Canadians... "Japanese-Canadian Applicant to Build Mill," *FVR*, November 11, 1937, 1; Dore, *From Harbour to Harvest*.
Most Japanese immigrants to Mission... Dore, *From Harbour to Harvest*, 57–58.
Mill owners preferred new immigrants from the Punjab... "Kuldip Singh Bains," *100 Year Journey* (Surrey, BC: Mehfil Magazine, 2014), 85.
Indar Singh Gill... "Indar Singh Gill," *100 Year Journey* (Surrey, BC: Mehfil Magazine, 2014), 75; Khalsa Diwan Society (Abbotsford), "Indar Singh Gill," interview with Nash Gill, son of Indar Singh Gill, *Canadian Sikh Heritage*, canadiansikhheritage.ca.
Gill also served with Mission's East Indian Community Association. Untitled photo of Mission East Indian Community Association meeting with Mission's mayor and reeve, *FVR*, October 27, 1965, sec. 2, p. 1; Khalsa Diwan Society (Abbotsford), "Indar Singh Gill," *100 Year Journey*; "Rebuilding Sawmill, Doubling Capacity," *FVR*, June 22, 1960, 1.
... *Herman Singh Braich*... Mission Museum, "Shaping Our Community: Prominent Indo-Canadian Pioneers," missionmuseum.com; Susan Lazurak, "'Lion' Fought for Fellow Immigrants," *The Province*, June 28, 2009.
Naranjan Grewall... "Naranjan Singh Grewall," *100 Year Journey* (Surrey, BC: Mehfil Magazine, 2014), 142.

Grewall's mill was rebuilt, only to burn down again four years later. "Sawmill Not to Be Rebuilt in Mission," FVR, August 14, 1947, 8; "Fire Destroys Mission Sawmill," FVR, June 3, 1947, 1; "Fire Destroys Flats' Sawmill This Morning," FVR, June 27, 1951, 1; "Another Sawmill for City Work Starts, Plant Coming," FVR, August 10, 1955, 1.

Grewall used his business influence to press for political change... "Grewall Heads Village Vote; Dead-Heat Feature in District," FVR, December 20, 1950, 1; "Grewall Named as Chairman of Board," FVR, January 6, 1954, 1.

... likely the first elected official of South Asian descent... "He Deemed It a Privilege to Live and Work Here," FVR, July 17, 1957, sec. 2, p. 2.

5: The Case for Conservation

... the Mission and District Board of Trade invited E.T. Calvert... Mission and District Board of Trade President's Annual Report for 1943, Mission Community Archives.

... the board invited a second forester, F.S. McKinnon, to speak... "Provincial Forester Speaks at Board of Trade," FVR, June 15, 1944, sec. 1, p. 1.

What does a community forest offer to local citizens? "Community Forests and Their Place in British Columbia," FVR, June 22, 1944, sec. 1, p. 1.

... in July 1944, the Mission District's council unanimously passed a measure... Minutes of Mission District Council meeting, July 1944; "Making Progress with Community Forest Project," FVR, October 12, 1944, sec. 1, p. 1.

Recommendations from the resulting Fulton Report, issued in 1910, shaped the 1912 Forest Act... Peter H. Pearse, "Evolution of the Forest Tenure System in British Columbia," unpublished monograph, (Vancouver, BC, 1992), 15, gov.bc.ca.

It is at this point that the odd corners... "Community Forests and Their Place in British Columbia," FVR, June 22, 1944.

... the Sloan Commission recommended two new types of tenure. Pearse, "Evolution of the Forest Tenure System," 17–23.

... the companies soon captured much of the available higher-value timber supply... Hayter, *Flexible Crossroads*, 48.

... "community forests, apart from the timber production therefrom..." Government of BC, *Public Inquiries Act—Report of the Commissioner relating to The Forest Resources of British Columbia*, 147.

... board of trade president A.G. McInnes called for "an inventory of the forest resources..." "Community Forests and Their Place in British Columbia," FVR, June 22, 1944.

E.T. Calvert was hired... "Mission Municipal Forest Reserve Bylaw Now Under Consideration," FVR, November 11, 1948, sec. 1, p. 1.

... "the Council of any Municipality may by by-law set aside..." District of Mission, *Brief, submitted to the Royal Commission on Forestry in British Columbia, by the Corporation of the District of Mission*, January 3, 1956, Mission Community Archives, FOR1-002.

... "Mission Municipality may one day have a similar experience..." Quote from broadcast was repeated by Reeve A.B. Catherwood in an interview printed in the *Fraser Valley Record*, "Value of Forest Reserve Bylaw Outlined by Municipal Council," December 2, 1948, sec. 1, p. 1.

More than 90 percent of the 672 votes cast were in favour... "Thompson and Hill Win Council Election; Forest Bylaw Endorsed," FVR, December 23, 1948, sec. 1, p. 1; "Mission Municipal Forest Reserve Bylaw Now Under Consideration"; District of Mission, *Brief, submitted to the Royal Commission on Forestry in British Columbia*.

... *the reserve allowed the District to experiment*... "3500 Fir Trees Are Planted in Local Experiment," FVR, March 23, 1955, sec. 1, p. 1.

... *small operators complained that they were shut out of the system*... Pearse, "Evolution of the Forest Tenure System," 19.

... *the problem of speculation*... "Associated Boards Back Logging Operator Issue," FVR, December 5, 1951, sec. 2, p. 3.

W.F. Watkins, owner of Watkins Sawmills, called for the creation of an appeals board... "Local Interest in Forest Policy Emphasized before Commission," FVR, February 2, 1956, sec. 1, p. 1. Article reproduces a brief submitted to the second Royal Commission on Forestry, on behalf of AA&A Cameron Sawmill, Savage Lumber Co. Ltd., Thomas Lumber Co. Ltd., and Watkins Sawmills Ltd.

... *"Timber Maharajahs"*... "Naranjan Grewall Warns against Rising Maharajahs and Feudalism," FVR, June 8, 1955, 1. Includes full text of Grewall's submission to the Forestry Commission (the second Sloan commission).

Grewall "has consistently fought for the rights of the small logging and sawmill operators." Dewdney CCF Council, "May We Introduce N.S. Grewall, Dewdney CCF Candidate," advertisement, FVR, August 22, 1956.

... *overly conservative estimates of timber supply*... "BC Lumber Industry Hinges on Proper Management, Utilization," FVR, January 21, 1953, sec. 1, p. 1.

... *investments in replanting*... "Local Interest in Forest Policy Emphasized before Commission," FVR, February 2, 1956.

... *"it takes 35 to 40 men..."* "Assures Permanence Lumbering Industry," FVR, March 21, 1951, sec. 1.

... *invariably have working agreements with coast sawmills*... "Local Interest in Forest Policy Emphasized before Commission," FVR, February 2, 1956.

McMahon referred to Mission's forest sector as a "hemlock economy." "Foresees Changes in Forest Industry," FVR, April 7, 1965, sec. 1, p. 1; "Indefinite Closure for Mission Mill," FVR, April 21, 1965, sec. 1, p. 1.

... *Mission's remaining sawmills were in a free fall.* McCombs and Chittenden, *Fraser Valley Challenge*, 77–78. Thomas's mill was irreparably damaged by fire in 1963.

... *Anglo American Cedar Products Ltd.*... McCombs and Chittenden, *Fraser Valley Challenge*, 79.

Founding members of the MSFA... Membership lists have been compiled from two reports within the *Fraser Valley Record*: "School Forest Plan Awaits Management License OK," March 17, 1954; "Community Forest Pattern—Set by Mission Program," May 25, 1955.

... *the MSFA identified, mapped, and moved onto thirty-five acres*... "School Forest Association Gets 35 Acres for Course", FVR, June 1, 1955, sec. 1, p. 1; "Mission Looking Ahead," *The Province*, March 31, 1964.

... *a newly developed forestry course*... "Mission Municipal Forest License, School Forestry Course, Lead B.C.," *Vancouver Province*, June 29, 1955, 12.

In the classroom and in the field, the students... "School Forest Plan Awaits Management License OK," *FVR*, March 17, 1954, sec. 1, p. 1.

... *management of nearby Crown lands*... Peter Grant, Chair, Forestry Committee of the Mission & District Board of Trade, "Re: Establishment of a Community Forest," letter addressed to the Reeve & Council of Mission Municipality, July 19, 1944. BC Archives, File GR-0520.17.8.8—Exhibit 423.

In 1955, Mission applied for a tree farm licence... District of Mission, *Brief, submitted to the Royal Commission on Forestry in British Columbia.*

The licence was designed primarily for large logging companies... Richard Rajala, "'Nonsensical and a Contradiction in Terms': Multiple-Use Forestry, Clearcutting and the Politics of Fish Habitat in British Columbia, 1945–1970," *BC Studies* 183 (2014), 90–91; Hayter, *Flexible Crossroads*, 49.

The province instead encouraged local harvesters and communities to join together under a public working circle. Rajala, "'Nonsensical and a Contradiction in Terms,'" 91; Hayter, *Flexible Crossroads*, 50.

... *public working circles in the Fraser Valley often failed to materialize*... "Board of Trade Endorses Battle Against Mount Sumas Control," *FVR*, January 9, 1952, 1.

Alf Buckland, Mission's municipal forester, recommended in 1953 that a new management plan be developed... "Management Plan Municipal Timber Being Considered," *FVR*, February 18, 1953, sec. 1, p. 1.

The District and the province reached an agreement-in-principle in 1954... Allan and Frank, "Community Forests in British Columbia," 721.

... *the Ministry of Forests put Mission's application for a TFL on hold*... R.E. Summers, Honourable Minister of Lands and Forests, letter to W. Matheson, Reeve of the Corporation of the District of Mission, December 5, 1955, District of Mission Forestry Department records.

Sloan's report, issued in 1957... Government of BC, *Public Inquiries Act—Report of the Commissioner relating to The Forest Resources of British Columbia*, 1956 (2 vols.) The Honorable Gordon McG. Sloan, Commissioner (Victoria, BC, 1956), 742–52.

... *"for at least twenty-five years all money earned..."* District of Mission, *Brief, submitted to the Royal Commission on Forestry in British Columbia*, 9.

6: A Growing Operation

... *Robinson managed a crew of five men*... "Perpetuation Aim of BC Forest Service," *FVR*, August 2, 1950, 1.

Working under him was Rocky Rockwell... "Perpetuation Aim of BC Forest Service," *FVR*, August 2, 1950. In addition to Robinson and Rockwell, the other crew members were C.J. "Bud" Wagner, Doug Brewis, and dispatcher Stan Deane.

The two agreed on the spot to split the ten-acre parcel... Lorne Rockwell, interview with Jillian Wheatley, 2017.

... *in November 1958, the Mission council named Robinson to be Buckland's replacement*... "Council Names Robinson Tree Farm Ranger," *FVR*, November 26, 1959, sec. 1, p. 1.

Problems of access... The Corporation of the District of Mission, *Tree Farm License No. 26—Annual Report (1959)*, January 15, 1960, physical files of the District of Mission Forestry Department.

... *Mission's barebones forestry team*... "Mission Tree Farm—Labour of Love for 20 Years," FVR, December 28, 1978, sec. 1, p. 1.

Speaking before the BC Legislature in 1959... "Municipal Tree Farm Commended to Others," FVR, February 11, 1959, sec. 2, p. 6

Three of the four cutting permits awarded... "Local Tree Farm Timber, Salvage Permits Awarded," FVR, November 19, 1958, sec. 1, p. 1.

Schedule A and B lands were managed together... Clause 18, Mission Municipal Tree Farm Licence, Tree Farm Licence No. 26, p. 6, digital files of Mission Forestry Department.

In 1958, most of Mission's forest could support some commercial timber extraction... Davis McCarey, BC Forest Service, memo to Mr. Holyhead, subject: Mission Municipal Tree Farm Tenure, August 4, 1958, digital files of Mission Forestry Department; The Corporation of the District of Mission, *Annual Report, 1960*, physical files of Mission Forestry Department.

Once a year, cruised blocks of timber were put out for tender... Scott Honeyman, "100-Year Wooden Dollars Mission Plan for Future," *Vancouver Sun*, July 22, 1968.

Up to 90 percent of the tenure burned... R.R. Lafferty, "Regeneration and Plant Succession as Related to Fire Intensity on Clear-Cut Logged Areas in Coastal-Cedar Hemlock Type," internal report BC-33, produced for the Pacific Forest Research Centre, May 1972, unpublished.

In 1938, almost seventy-five thousand acres burned... BC Fire Service, "Fire: The Good Servant" (film), 1964, digitally remastered by the University of British Columbia.

"The public should be informed that fire is the number one threat..." "Loggers Agenda Includes Forest Management," FVR, January 16, 1952, sec. 2, p. 1.

... *North America's first Junior Forest Warden*... "6500 Boys Active in CFA Junior Warden Movement," FVR, May 23, 1956, sec. 2, p. 4.

... *"dryer than dry"*... "Municipal and BCFS Crews Battle Fires," FVR, July 30, 1958, sec. 1, p. 1.

... *one at the head of Stave Lake in 1960*... "Million Feet Stave River Timber Burns," FVR, July 1, 1960, sec. 1, p. 1; "Difficulty Forest Fire Now Doused," FVR, September 7, 1960, sec. 1, p. 1; "Stave Lake Forest Fire Caused $200,000 Damage," FVR, November 2, 1960, sec. 1, p. 3.

"We know that they can't go on forever..." Honeyman, "100-Year Wooden Dollars Mission Plan for Future."

Revenues were used... "Perpetual Crop Production Plan," FVR, July 7, 1960, spec. sec., p. 1.

... *the District council had already spent nearly ten thousand dollars on an experiment*... Symonds, "Canada's Only Municipal Tree Farm."

One of those expressing skepticism was F.R. Hall... "$100,000 Tree Farm Revenue," FVR, July 22, 1964, spec. sec., p. 1.

"Last winter we were able to practically use all unemployed persons..." "Tree Farm Plan Outstanding," FVR, September 2, 1964, sec. 2, p. 2; "Tree Farm Project Seen as Boost to Economy," FVR, August 26, 1964.

... *in 1970, prices dropped so low*... "Second Chance for Tree Farm Bidders," FVR, January 28, 1970, sec. 1, p. 1; "Big Drop in Bids for Municipal Timber Sales," FVR, March 11, 1970, sec. 1, p. 1.

... *the resulting savings averaged fifty thousand dollars*... "Mission Tree Farm—Labor of Love," FVR, December 28, 1978, sec. 1, p. 1.

Generations of settlers had toiled... "Tree Farm Plan Outstanding," FVR, September 2, 1964; "Tree Farm Project Seen as Boost to Economy," FVR, August 26, 1964.
In early 1944, the District had attempted a land purchase... Symonds, "Canada's Only Municipal Tree Farm."
... *"make the area more suitable for recreation..."* "Garibaldi Timber Sought by Reserve," FVR, October 14, 1964, sec. 1, p. 1.
The timber berths in question were originally lands granted to the CPR... "Needs More Land," FVR, July 18, 1973, sec. 1, p. 1; "Tree Farm Is Given Big Boost in Acreage," FVR, August 7, 1974, sec. 1, p. 1.
"The short-term value is..." "Tree Farm Is Given Big Boost in Acreage," FVR, August 7, 1974.
"Culverts go where they should..." "Tree Farms in Vancouver," courtesy of the editor, *Vancouver Sun*. Reproduced in "Around the Commonwealth," in the *Commonwealth Forestry Review* 53, no. 3 (1974), 184–88.
... *another forty kilometres of road would be needed*... "Look to Future of Tree Farm," FVR, September 4, 1968, sec. 1, p. 1.
"For the first time since the tree farm license was granted..." "Sales Reach $300,000," FVR, November 26, 1968, sec. 1, p. 1.
Road building accelerated... Photos of Howard Murdoch, Mission Tree Farm Foreman, leading Mission Council members on a tour of road building projects, FVR, August 1, 1970, sec. 1, p. 1; "Plan $100,000 Outlay in Tree Farm," FVR, March 24, 1971, sec. 1, p. 1; "Mission Tree Farm... a Model for Others," FVR, January 9, 1974, sec. 1.
Site preparation for the WSAR... Photographs documenting different stages of the road's development were reproduced in *The Fraser Valley Record* on June 5 (sec. 3, p. 5), June 12 (sec. 3, p. 1), and November 13 (sec. 1, p. 7), 1963.
... *crews began erecting a permanent fifteen-metre-long bridge*... "Bridge to Span Canyon," FVR, July 20, 1966, sec. 1, p. 1.
... *the District auctioned off 210,000 board feet*... Legal advertisement for timber sales by the District of Mission, printed in the *Fraser Valley Record*, February 22, 1967, sec. 1, p. 7.
... *crews moved into forest beyond*... Symonds, "Canada's Only Municipal Tree Farm."

7: Branching Out

"Rocky Rockwell wasn't really an office guy..." Graham Webster, personal communication, September 2018.
"My dad used to get in the office in the morning..." Lorne Rockwell, interview with Jillian Wheatley, 2017.
Japanese importers sought out yellow cedar supplies... Lauren Oakes, *In Search of the Canary Tree: The Story of a Scientist, a Cypress, and a Changing World* (Basic Books, 2018).
Many of the trees were massive... "Visitors Impressed on Tree Farm Tour," FVR, June 19, 1968, sec. 1, p. 1.
The decision to begin working with yellow cedar... "Trial & Error Spells Success," *British Columbia Lumberman*, September 1978, 48.
"Instead of waiting for someone else..." "Tree Farm in the Spotlight," FVR, August 24, 1977, sec. 1.

... yellow cedar required a more complicated regimen... Moira Farrow, "Blast of Cold Air Fools Tree Seeds into Sprouting," *Vancouver Sun*, April 1, 1980, C1.
"Last year we ended up with only 800 saplings..." "Mission Tree Farm Starts Research," *FVR*, August 21, 1974, sec. 1.
Mission's nursery crews experimented... Data on nursery numbers found in the District of Mission Tree Farm License #26 Annual Reports; acreage figure from "Tree Farm in the Spotlight," *FVR*, August 24, 1977.
... yellow cedar survival rates... Farrow, "Blast of Cold Air Fools Tree Seeds."
After "much discussion"... The Corporation of the District of Mission, *Annual Report*, 1968, District of Mission Forestry Office files.
"Nobody was willing to take the responsibility..." "Mission Tree Farm—Labor of Love for 20 Years," *FVR*, December 28, 1979, sec. 1, p. 1.
Lafferty identified the tenure's "broken topography..." Lafferty, "Regeneration and Plant Succession," 8.
... Mission's milder climate permitted controlled burns... Canadian Forestry Association of British Columbia, Forest Fire Prevention and Control Group, *Fire Control Notes 1979*, 14th fire control course, Howard Murdoch folio, Mission Community Archives.
The project team carried out twenty-two burns... Lafferty, "Regeneration and Plant Succession," 67; "Study Air Pollution at Site of Slash Burning," *FVR*, July 31, 1968, sec. 2, p. 1; "Under Expert Control, Even Forest Fire Can Be Useful," *FVR*, August 12, 1970, sec. 1, p. 1.; The Corporation of the District of Mission, *Annual Report*, 1968.
... Lafferty continued collecting data... R.R. Lafferty, "Post-Burn Evaluation of Plant Succession and Forest Regeneration in Coastal Douglas Fir-Western Hemlock-Red Cedar Types," research proposal submitted on behalf of Havco Resource Services Co. to the Department of Supply and Services (Canada), 1980; M.P. Curran, "Slashburning Effects on Tree Growth and Nutrient Levels at Mission Tree Farm: Project Status" (poster), 1988; J.D. Lousier and G.W. Still, ed., *Degradation of Forested Lands: "Forest Soils at Risk": Proceedings of the 10th BC Soil Science Workshop, February 1986*, 294–313.
... a pair of biologists from Simon Fraser University... T.J.D. VanderSar and J.H. Borden, "Aspects of Host Selection Behaviour of *Pissodes strobi* (Coleoptera: Curculionidae) as Revealed in Laboratory Feeding Bioassays," *Canadian Journal of Zoology* 55 (1977), 405–14.
... "a long ways from nowhere"... "Marijuana in the Tree Farm," *FVR*, February 6, 1974, sec. 1.
... forestry and wilderness survival program at Boulder Bay... BC Corrections Branch History Committee, "History of BC Corrections Camps 5" (video), video slideshow about Boulder Bay Camp narrated by Fred Hunt, gov.bc.ca; Government of BC, Law, Crime & Justice, "1950s: First Correctional Forestry Camps Established, Including the Oakalla Camps," gov.bc.ca. The Boulder Bay facility closed in 2002.
Over several summers, BC Hydro drew down the reservoir... Government of BC, Law, Crime & Justice, "1971: Stave Lake Camp / Correctional Centre," gov.bc.ca.
By the end of 1976, they had cleared more than thirty acres. "At Stave Camps Hard Work Earns Credit," *FVR*, September 19, 1976, sec. 1.
Inmates tended to injured birds... Heather Kent, "Flying the Coop," *Canadian Geographic* 118, no. 6 (September/October 1998); "Prisoners Treat Nature Well," *Mission City Record*, September 3, 1998, 22.

BC Corrections also ran a second short-lived camp... "At Stave Camps Hard Work Earns Credit," *FVR*, September 19, 1976.

The camp met its demise in the mid-1990s... "Former Prison Camp Destroyed," *Mission City Record*, February 23, 1994.

... *"environmental, education and community recreation purposes only."* "Work Camp Turned Over to Area Schools," *FVR*, April 18, 1979, sec. 1.; "Sayers [sic] Lake Camp Is Up for Sale Again, for Only $1," *FVR*, March 16, 1981, sec. 1.

The District of Mission leased the whole site for one dollar to SLCS... John Alexander, "Potential of Cedar Camp Shown by Early Usage," *FVR*, August 15, 1979; "Sayers [sic] Lake Camp Is Up for Sale Again," *FVR*, March 16, 1981.

... *"massive assault on the province's forests."* British Columbia, *Report of the Royal Commission on Forest Resources—Timber Rights and Forest Policy in British Columbia*, Peter H. Pearse, Commissioner (Victoria: Queen's Press, 1976), 4.

The commission believed that the "success of the Tree-farm Licence..." British Columbia, *Report of the Royal Commission on Forest Resources*, 118.

... *"a rather arbitrary transfer of revenue..."* British Columbia, *Report of the Royal Commission on Forest Resources*, 119.

"We don't think it's fair..." Farrow, "Blast of Cold Air Fools Tree Seeds."

The Mission Tree Farm raked in surpluses... District of Mission, *Tree Farm License #26 Annual Report 1986*, District of Mission Forestry Department files.

8: A Community's Forest

... *reserve fund specifically for community arts and culture...* Scott Simpson, "Mission Leads Municipal Logging Trend," *Vancouver Sun*, November 22, 1997, B1.

Tower logging... A.J. McDonald, ed., "Harvesting Systems and Equipment in British Columbia," handbook prepared by the BC Ministry of Forests, Forest Practices Branch (1999), gov.bc.ca.

"When the Mission Tree Farm emerged out of nowhere..." Norm Horn, interview with Jillian Wheatley, District of Mission practicum student, and George Kocsis, July 2017.

... *a local logger, Dale Soper, used a horse-powered skyline system...* Glen Kask, "Horse Logging in Mission's Forest," *FVR*, October 16, 1991, 1.

The BC Forest Service and Parks Branch cleared the land on the eastern edge... "Development Soon Two Local Lakes," *FVR*, May 11, 1960, sec. 1, p. 1; W.R. Jack, "From the District: Rolley Lake Park" *FVR*, June 7, 1961, sec. 2, p. 2.

... *"a popular picnic spot..."* "Development Soon Two Local Lakes," *FVR*, May 11, 1960.

... *did not eliminate all unauthorized trails...* Glen Kirk, "Trails Torn Up by City," *Mission City Record*, September 30, 1999, 1.

... *Zajac Ranch...* Kim Pemberton, "Kids Camp Planned for Old Prison," *Vancouver Sun*, September 25, 2003, B2.

... *Tim Horton's project...* Jason Roessle, "Camp Would See $15 M Invested," *Mission City Record*, September 24, 2012, missioncityrecord.com; District of Mission Forestry Department, *Stave West Master Plan*, 2015, 8, 135.

As of 2018, an additional $3.5 million was needed... District of Mission Forestry Department, *Stave West Master Plan*, 80.

By the mid-2000s, the District was soliciting feedback from several more Nations... District of Mission, *Mission Municipal Forest Sustainable Forest Management Plan for Mission Tree Farm Licence #26*, March 31, 2003, mission.ca; District of Mission, *Tree Farm Licence No. 26 (Mission Municipal Forest) Forest Stewardship Plan, 2007–2012*, Amendment #1, May 27, 2011, mission.ca.

9: The Wild Stave West

... *eleven abandoned vehicles over a twenty-kilometre stretch*... Amanda McCormick, Daryl Plecas, and Irwin Cohen, *Motor Vehicle Theft: An Analysis of Recovered Vehicles in the Fraser Valley* (School of Criminology and Criminal Justice, University College of the Fraser Valley, November 2007), ufv.ca.; Carol Aun, "More Patrols in Recreational Areas," *Mission City Record*, May 18, 2006, 1.

Mission further expanded the monitoring program in 1999. Cheryl Wierda, "Crime Prevention Program at Stave Lake Expands," *Mission City Record*, April 26, 2001, 1; District of Mission Forestry Department, *Annual Reports*, 1995–2003.

"Only the District of Mission seems to want to do something..." Roxanne Hooper and Cheryl Aun, "Shores of Stave Lake a Dumping Ground," *Mission City Record*, June 5, 2003, 1.

... *the untapped potential for tourism*... Marcia Downham, "Trashed Woods Could Be Outdoor Haven," *Abbotsford Times*, August 12, 2008, A4.

... *recreation and tourism feasibility study*... District of Mission, *TFL 26 Recreational Opportunities Feasibility Analysis*, May 5, 2009, prepared by LEES + Associates Landscape Architects with G.P. Rollo and Associates Land Economists Murray Management, Inc., mission.ca.

... *ten guiding principles*... District of Mission Forestry Department, *Stave West Master Plan.*

... *the Kwantlen received their licence for the woodlot on Blue Mountain*... Government of BC, Kwantlen First Nation Interim Agreement on Forest Opportunities (March 22, 2006), gov.bc.ca; Phil Melnychuk, "New Woodlot for Blue Mountain Approved," *Maple Ridge News*, December 7, 2011.

The K&K includes 6,732 hectares of forest... səyem̓ qwantlen business group, SQ / Council updates for October 5, 2017.

Approximately 21 percent of what was harvested in 2018... Ministry of Forests, Lands, Natural Resource Operations, and Rural Development, Report: TSA, AAC and Commitments, Fraser TSA, April 6, 2018, gov.bc.ca.

The Leq'á:mel have given notice to the province and to Mission... Leq'á:mel First Nation, Assertion of Title, March 8, 2017, leqamel.ca.

photo: Jason Brawn

ABOUT THE AUTHOR

A GEOGRAPHER BY TRAINING, Michelle Rhodes is interested in how small communities change over time in response to significant economic and environmental challenges. She is director of Integrated and General Studies and an associate professor of geography at the University of the Fraser Valley. Her previous research has explored the economic viability of community forestry and the rise of mass-produced housing in small towns. Michelle lives in Mission with her husband and two dogs. She is grateful to live in the shared territory of the Stó:lō Peoples, and in particular the Kwantlen, Matsqui, and Leq'á:mel, and is thankful for the bounty and beauty that the land and waters of this region provide.